CW01279000

GNVQ Core Skills

APPLICATION OF NUMBER

Intermediate/Advanced

SECOND EDITION

Catriona Johnstone

Lecturer in Application of Number and a GNVQ Course Manager
Farnborough College of Technology

PITMAN PUBLISHING

PITMAN PUBLISHING
128 Long Acre, London WC2E 9AN

A Division of Pearson Professional Limited

First published in Great Britain in 1994
Second edition 1995

© Catriona Johnstone 1994, 1995

The right of Catriona Johnstone to be identified as Author
of this Work has been asserted by her in accordance
with the Copyright, Designs and Patents Act 1988.

ISBN 0 273 62053 3

British Library Cataloguing in Publication Data
A CIP catalogue record for this book can be obtained from the British Library.

10 9 8 7 6 5 4 3 2 1

Typeset by M Rules
Printed and bound in Great Britain by Clays Ltd, St Ives plc

The Publishers' policy is to use paper manufactured from sustainable forests.

Contents

Introduction

The Application of Number core skill unit has three elements:

- Collect and record data (Element 1)
- Tackle problems (Element 2)
- Interpret and present data (Element 3)

and each element identifies three key themes:

- Number
- Shape, space and measures
- Handling data

The aim of GNVQ Application of Number is to help the candidate understand and apply, in a vocational setting, much of the mathematics learnt at school. For many this book will only be used for revision and as an indication of what is required as evidence. For others it will be an essential study aid to learn the mathematical techniques required to obtain a GNVQ qualification.

Before you begin learning the mathematical techniques you should understand how the system works. The Application of Number core skill consists of three **elements**. Each element has **performance criteria** and **range**. The range contains the **mathematical techniques** you need to demonstrate (the mathematics questions you need to answer correctly). The **amplification** of the element identifies the performance criteria to be fulfilled for the various mathematical techniques.

The three complete elements for level 2 are:

■ Element 2.1: Collect and record data

The student decides which technique to use in a range of simple data-collection and measuring tasks which they have designed themselves. The student decides which level of accuracy and tolerances to use but is not expected to determine the sample size for the data-collection task.

PERFORMANCE CRITERIA

A student must:

1   make decisions about what data should be collected
2   choose and use **techniques** which suit the task
3   perform the **techniques** in a correct order
4   choose and work to an appropriate **level of accuracy**
5   record data in appropriate **units** and in an appropriate format
6   make sure that records are accurate and complete

RANGE

Techniques:

- **Number:** describing situations (using fractions, using decimal fractions, using percentages, using ratios, using negative numbers); using estimation (to judge quantities, to judge proportions, to check results, to predict outcomes)
- **Shape, space and measures:** choosing and using appropriate measuring instruments and appropriate units of measurement for the task
- **Handling data:** designing and using a data collection procedure for a given sample; obtaining data from written sources, obtaining data from people; working with discrete and grouped data.

Levels of accuracy: choosing and using an appropriate level of accuracy for each measuring task, choosing and working within appropriate tolerances for each measuring task

Units: of money, of physical dimensions, of another property

EVIDENCE INDICATORS

Separate reports by the student of at least two data collection tasks where the student has designed the data collection procedures. The reports should include reasons for: collecting the data, choosing the methods for collecting the data, and choosing the means of recording the data. These two reports could also provide evidence for the other aspects of data collection and recording which are listed in the following paragraphs.

Data collection sheets designed by the student which display the results of the following data collection and measuring tasks:

- at least two tasks involving money
- at least two tasks involving measurement of physical dimensions
- at least two tasks involving data collection/measurement of one other problem.

These reports should include evidence of extraction of data from at least two written sources and from at least two tasks involving collection of data from people.

Records or responses to oral questions put to the student which show the student's capacity to estimate quantities in relation to distance and size; at least two tasks to be completed for each of these.

Notes of working should show the student's capacity to work in a range of settings with a range of numbers (eg rounding to whole numbers; working to three decimal places), to understand and work with tolerances, and to use estimation to check results and predict outcomes.

Skills in this element should be learned and demonstrated through activities which will enhance the student's capacity to perform effectively in vocational settings.

AMPLIFICATION

Choose and use techniques (PC2): at this level, the student should design data collection and measuring procedures and decide on appropriate levels of accuracy (eg deciding what questions to ask to determine residents' reactions to increasing numbers of visitors to a National Park; designing a data collection sheet to record clearly and accurately the quality of engineering products coming off a production line). At this level, the student is not expected to determine the sample her/himself (eg the number of people to ask about an issue; the number of products to measure).

Fractions (PC2 and PC3 range): examples include about $\frac{2}{3}$ (eg to describe the approximate proportion of customers spending over £50 in a shop), over $\frac{3}{4}$ (eg to describe the approximate proportion of telephone calls made from an office which are long distance and made during peak charge times).

Decimal fractions (PC2 and PC3 range): examples include 0.85m (eg when measuring the dimensions of an object), 0.35 l (eg when measuring out fluids).

Percentages (PC2 and PC3 range): examples include under 5% (eg to describe the approximate proportion of time a computer network is failing to work), over 80% (eg to describe the approximate proportion of people visiting a National Park during the winter period at weekends rather than weekdays).

Ratios (PC2 and PC3 range): examples include 2:5 (eg to describe two materials mixed together), 1:15 (eg to describe the average teacher-student ratio in teaching sessions in college departments).

Negative numbers (PC2 and PC3 range): examples include –0.5°C (eg when measuring daytime winter temperature), –£20,000 (eg when looking at end-of-year trading losses for a company).

Estimation to judge quantities and proportions (PC1 and PC2 range): the student should be able to estimate quantities and size when collecting data (eg 'this corridor is about 8.5 metres long'; 'if we stack up six of those boxes they will come to over 2m'; 'each file is just over 2cm thick'), and estimate fractions (eg 'that's about a quarter') and percentages (eg 'that's about 30%').

Estimation to check results and predict outcomes (PC2 and PC3 range): this should be used by the student to check whether results are reasonable, using knowledge of the context and/or the size of numbers which might be expected.

Measuring instruments (PC2 and PC3 range): these should be selected by the student in order to collect the right sort of data at the level of accuracy which s/he has decided is appropriate to the task (eg a thermometer which measures to a tenth of a degree; a steel rule with mm marked; weighing scales with grams marked).

Data collection procedure (PC2 and PC3 range): the emphasis is on designing valid and effective procedures for exploring problems and questions; the precise procedures can include using a questionnaire (eg of people's reactions to the level of service an organisation provides); a survey using observation (eg the routes people take through a building); or an experiment (eg how a material responds to changes in temperature).

From written sources (PC1 and PC2 range): examples include paper-based materials (eg a list of annual sales figures in a company's annual report; a table in a newspaper displaying the population of different towns) and IT-based materials (eg a chart from a database showing the temperature variation over a year; figures from a spreadsheet on annual profits).

From people (PC1 and PC2 range): examples include asking a sample of people how much they spend on specific products; asking people the distance they travel, the time they take and the amount it costs.

Discrete and grouped data (PC2 and PC3 range): collecting and recording discrete data (eg numbers of children per couple), grouping data (eg measuring temperature changes to the nearest degree, at intervals of 12 hours and drawing up a frequency table of the results).

Levels of accuracy (PC4): at this level, the student should select and work to a suitable level of accuracy for data collection and measuring procedures which s/he has designed (eg be able to round to the nearest decimal place in weighing a range of engineering components; work to three decimal places in working with data from measuring tasks), and be able to work with numbers of any size (eg numbers of cars in the UK at 23,000,000; internal bore of an hypodermic needle at 0.04mm).

Tolerances (PC4 range): the student should be able to set and work within tolerances (eg 5mm for a task measuring dimensions; 15min timing an activity).

Units of physical dimensions (PC5 range): the student can work with either metric or imperial units. Examples include measuring width, depth and height of items, measuring distance between locations or objects. Note that in Element 2.2 the student will be expected to convert from one system to another.

Units of another property (PC5 range): examples include temperature (°F, °C), time (hours, minutes, seconds), metric units of capacity (l, ml), metric units of mass (kg, g), imperial units of mass (lb, oz).

GUIDANCE

The tasks should be related to the settings in which students are operating. Examples of suitable opportunities include:

- designing a questionnaire and data collection sheet to ask 15 employees about their hours of work and rates of pay, carrying out the survey and recording the results to an appropriate level of accuracy
- estimating and then measuring and recording the temperature in an office at set times of the day over a month to see how this compares with health and safety recommendations.

■ Element 2.2: Tackle problems

The student decides which techniques to use to solve problems. In each case, solving the problem should involve using a series of techniques with numbers of any size. The student uses both calculator and non-calculator methods to make calculations correctly and to check the results. The student decides which level of accuracy to use.

PERFORMANCE CRITERIA

A student must:

1 choose and use **techniques** which suit the problem
2 perform the **techniques** in a correct order
3 choose and use appropriate units
4 choose and use an appropriate **level of accuracy**
5 use mathematical terms correctly
6 carry out calculations correctly

7 use **checking procedures** to confirm the results of calculations

8 check that the results make sense in respect of the problem being tackled

RANGE

Techniques:

- **Number**: working with numbers of any size (using addition, using subtraction, using multiplication, using division); calculating (with fractions, with decimal fractions, with percentage, with ratios); calculations using simple formulas (expressed in words, expressed in symbols)
- **Shape, space and measures**: calculations using common units of measurement; solving problems in two dimensions (perimeters of shapes, the areas of plane shapes), solving problems in three dimensions (volume of simple shapes, volume of cylinders)
- **Handling data**: converting between common units of measurement using scales and tables, converting using calculations; calculating and using mean, median, mode and range

Levels of accuracy: selecting and using an appropriate level of accuracy; approximations

Units: approximate calculations, estimating, inverse operations

EVIDENCE INDICATORS

Work in the form of notes, sections of projects or reports, and/or workbooks, which shows detailed working in the procedures which the student used to tackle problems and tasks, and gives the reasons why particular approaches were chosen. The results of calculations should be accurate, and evidence of checking should be present. Adequate evidence should be presented to show that the student is confident in the routine use of each technique listed in the range. This will mean the presentation of evidence which shows working with each technique in at least two tasks or problems.

The work should show that the student is able to work to different levels of accuracy in a range of settings, as demanded by the problems and tasks being tackled and as described in the range and amplification.

The skills in this element should be learned and demonstrated through activities which will enhance the student's capacity to perform effectively in vocational settings.

AMPLIFICATION

Choose and use techniques (PC1): the student should be able to make decisions about which technique to use in response to particular problems and tasks. At this level, the student is expected to deal with problems which demand use of a series of techniques (eg determining the use of materials by different teams on a construction site over a one-month period, requiring: the data to be divided into groups; converting some data from one unit to another; calculating means for each group of data and then comparing them). At this level, the student should be able to use all four operations (+ − × ÷) with decimals of any size, calculate fractional/percentage parts of quantities and measurements using a calculator where appropriate, and use non-calculator methods for multiplying/dividing any three-digit number by any two-digit number.

Numbers of any size (PC1 and PC2 range): the student should be able to work with positive numbers of any size (eg 2,377,060; 2.79; 0.003); s/he should be able to work with negative as well as positive numbers (eg working out the total profit/loss over a five-year period for a company which repeatedly lost money over three years but then went into profit for the last two years).

Fractions (PC1 and PC2 range): calculations with fractions such as $\frac{1}{4}$, $\frac{2}{5}$. At this level, students should be able to work effectively with fractions when solving problems (eg be aware that $\frac{3}{7}$ is larger than $\frac{3}{10}$ when working out quantities of materials).

Decimal fractions (PC1 and PC2 range): calculations with decimal fractions (eg 0.25m, 0.4l).

Percentages (PC1 and PC2 range): calculations with percentages (eg 23% of the cost of a product, 17.5% for the calculation of VAT) and calculation of proportional change (eg comparing output statistics for 1993 with those for 1994 show a 16% increase in output).

Ratios (PC1 and PC2 range): calculating and comparing different ratios (eg to compare different concentrations of fluids, where one liquid is 1:20 solution, another is 7:8).

Simple formulas expressed in words (PC1 and PC2 range): examples include total cost = cost of production + distribution + profit; volume of water = inflow minus loss through evaporation.

Simple formulas expressed in symbols (PC1 and PC2 range): examples include $V = I \times R$ (in electricity); $s = \frac{d}{t}$ (in calculating speed); $F = \frac{9}{5}C + 32$ (in converting data on temperature from Fahrenheit to Celsius).

Volume of simple shapes (PC1 and PC2 range): examples include calculating the volume of an L-shaped room in order to decide what size central-heating radiators should be used; calculating the volume of a rectangular tank and working out how much is in it when it is $\frac{1}{4}$ full.

Volume of cylinders (PC1 and PC2 range): the student should be able to calculate the volume of different cylinders and compare the variation in volumes (eg how the volume of a syringe 2cm in diameter and 8cm long compares with another which is also 8cm long but is 1cm in diameter).

Converting (PC1 and PC2 range): this should be completed in two ways: within single systems (eg 1.5m = 1500mm; and 2ft = 24in) and across systems (eg 1ft = 0.3m, and 37°C = 98.6°F). At this level, students should be able to use conversion constants (eg 0.1m equals 10cm) and formulas (eg to convert from °Celsius to °Fahrenheit).

Levels of accuracy (PC4): at this level the student is expected to select a suitable level of accuracy (eg rounding to whole numbers; working to three decimal places).

Approximations (PC4 range): examples include rounding to the nearest whole number, to the nearest 10.

Approximate calculations (PC7 range): for example, using whole number calculations which have been worked out using non-calculator methods in order to check whether the results of more precise calculations look correct.

Inverse operations (PC7 range): examples include checking subtraction by addition, and division by multiplication.

GUIDANCE

The techniques could be performed on the data collected for Element 2.1 and should be related to the settings in which the student is operating. Examples include:

- calculating the mean, median and mode for the hours of work and rates of pay of the employees surveyed in Element 2.1 and expressing these as both a decimal fraction and a percentage of the national average. Grouping the data into bands of pay (eg £180–£200 per week) and hours of work (eg 37.5–40 hours per week) and calculating the percentage of employees in each band
- calculating the mean, median and mode for the temperature of the office measured in Element 2.1 and identifying the maximum and minimum temperature for each day. Converting these results from °Celsius to °Fahrenheit.

Element 2.3: Interpret and present data

The student identifies the main features of the data and relates these to the problem tackled. The student decides which more complex techniques to use to display data and which titles and labels to use. The student decides what level of accuracy and which scales and axes to use.

PERFORMANCE CRITERIA

A student must:

1 identify and **explain the main features** of the data
2 choose and use **techniques** which will present the data effectively
3 follow **conventions** for presenting the data
4 present the results with an appropriate **level of accuracy**
5 explain how the results make sense in respect of the problem being tackled

RANGE

Explain the main features: expressing in words the main relationships and patterns, expressing relationships in symbols, expressing relationships involving rate

Techniques:

- **Number:** interpreting and presenting result of calculations; using probability to describe situations
- **Shape, space and measures:** interpreting and constructing two-dimensional diagrams (plans, drawings); interpreting and constructing two-dimensional representations of three-dimensional objects
- **Handling data:** interpreting and constructing statistical diagrams (pictograms, bar charts, pie charts), interpreting and constructing graphs; interpreting mode, interpreting mean, interpreting median, interpreting range

Conventions: selecting and using appropriate axes and labels

Levels of accuracy: selecting and using appropriate units for presenting data; selecting and using appropriate scales

EVIDENCE INDICATORS

Written presentations of data from written reports or spoken presentations, which display data using all of the different forms listed in the range for Shape, space and measures and Handling data. The presentation of data in this element can use the data collected in Element 2.1 and/or the results of calculations in Element 2.2.

Evidence should show that the student is confident in the use of each technique listed in the range. At least two examples of each form of diagram etc listed in Shape, space and measures and Handling data should be presented; these should demonstrate that the student is able to choose and use axes, scales and conventions appropriate to different settings.

For each piece of work, s/he should provide an explanation of the main features of the data and how these relate to the task or problem which is being tackled.

The skills in this element should be learned and demonstrated through activities which will enhance the student's capacity to perform effectively in vocational settings.

AMPLIFICATION

Explain the main features (PC1): this relates to the key features of a set of data (eg trends – such as the quantity of sales increasing over time; distribution – such as the way major industry is spread across a county; upper and lower limits – such as variation over a three-month period of temperature extremes in a building). It is essential that the student understands how the data relates to the problem being tackled (eg having monitored and displayed data on numbers of engineering components falling outside acceptable limits for different teams of workers, are some teams performing better over a one-month period than others? Do these monthly figures for the punctuality of trains mean that annual targets are being met?) At this level, the student should express the relationships and patterns in words (eg 'while turnover increased by 31% this year, operating profits were down by over 7%'), in symbols (eg 'profit equals income minus expenditure' expressed as $P = I - E$), and expressing relationships involving rate (eg income per month; profit per year; comparing death rate with birth rate over a five-year period).

Data (PC1): Elements 2.1 and 2.2 will produce data which can be used as the basis for the tables, charts and other forms of presentation which are described in this element.

Choose and use techniques (PC2): the student should make decisions about which is the most appropriate technique for displaying data for specific tasks (eg choosing an appropriate way to present a full set of data about production in a workshop in order to show clearly that 22.7% of a week's production run of 8,000 axles are exactly on the tolerance limit and that 17.2% are outside a tolerance of ±0.05mm) – see the range for Shape, space and measure and Handling data. The student should choose appropriate scales and axes for diagrams. At this level, the student should also choose the level of accuracy which would be most appropriate for presenting data (eg round to whole numbers, display to two decimal places) in order to show the key features of the data effectively.

Using probability to describe situations (PC2 range): numerical descriptions of the probability of events (eg 3% probability to say that 3 items out of every 100 from a production line are defective) and use of the probability scale as a way of describing the probability of events (eg 0.5 as a way of saying that something will probably happen 10 times out of 20).

Two-dimensional diagrams (PC2 range): the student should be able to construct plans (eg of an area in an engineering workshop, to show the position and orientation of five pieces of heavy equipment and the clearances around them), and drawings (eg to show the position of a set of switches at the back of a computer).

Two-dimensional representations of three-dimensional objects (PC2 range): the student should be able to construct drawings which convey the dimensions of objects (eg a sketch drawing of a laser printer showing its breadth, depth and height).

Statistical diagrams (PC2 range): the student should be able to construct pictograms (eg to show the distribution of different types of industry in a particular part of the country), bar charts (eg to show the numbers of customers in different age groups using a music store) and pie charts (eg to show the proportion of different types of equipment breakdowns over a two-month period).

Graphs (PC2 range): the student should be able to construct line graphs with more than one line which allow comparisons of different categories to be drawn and compared on the same graph (eg a graph which compares the growth rates of different types of service industry and manufacturing industry over a 20-year period).

Axes (PC3 range): the student should choose axes in order to show data to the best effect, ensuring that independent variables (eg time) are displayed along the bottom (x) axis.

Labels (PC3 range): accurate and clear labels should be attached to all scales, axes and all other components of d .ta.

Selecting and using appropriate scales (PC4 range): the student should be able to select scales which will allow important features of the data to be displayed effectively; this will include choosing scales of the appropriate intervals and units (eg deciding not to begin a scale at zero since the key data is temperature variation in a kiln between 900°C and 1200°C).

GUIDANCE

The data presented could be that collected in Element 2.1 and/or the results of the calculation of Element 2.2. Examples include:

- representing the spread of pay and hours of work of the employees surveyed in Element 2.1 in the form of bar charts and pie charts, using the percentages for certain bands calculated in Element 2.2. Relating the mean, mode and median values for pay and hours of work to the national average, explaining what the difference is and suggesting reasons why they differ
- using line graphs to show the spread of the temperature readings taken in an office over a month in Element 2.1; one line representing the maximum value for each day, one line representing the minimum value for each day and a line showing the mean value for each day. Comparing the results obtained with health and safety recommendations on office temperature and, if the results fall outside these recommendations, suggesting reasons why (eg poor air conditioning).

The three complete elements for level 3 are:

■ Element 3.1: Collect and record data

The student decides which technique to use in a range of data-collection and measuring tasks which s/he has designed her/himself, including determining an appropriate sample size. The student decides what level of accuracy and tolerances to use. The student also identifies sources of error and the impact of these errors on both large and small data sets.

PERFORMANCE CRITERIA

A student must:

1 make decisions about what data should be collected
2 choose and use **techniques** which suit the task
3 perform the **techniques** in a correct order
4 choose and work to an appropriate **level of accuracy**
5 record data in appropriate **units** and in an appropriate format
6 make sure that records are accurate and complete
7 identify sources of error and their effects

RANGE

Techniques:

- **Number:** describing situations (using fractions, using decimal fractions, using percentages, using ratios, using negative numbers); using estimation (to judge quantities, to judge proportions, to check results, to predict outcomes)
- **Shape, space and measures:** choosing and using appropriate measuring instruments and appropriate units of measurement for the task
- **Handling data:** selecting an appropriate sample and designing and using a data collection procedure; obtaining data from written sources, obtaining data from people; discrete data, continuous data; handling large data sets

Levels of accuracy: choosing and using an appropriate level of accuracy for each measuring task, choosing and working within appropriate tolerances for each measuring task

Units: of money, of physical dimensions, of another property, of rates of changes

EVIDENCE INDICATORS

Separate reports by the student of two data collection tasks where the student has designed the data collection procedures; one task should involve the student in managing large sets of data. The reports should include reasons for: collecting the data, choosing the methods for collecting the data, and choosing the means of recording the data. Each report should include the reasons for choosing a particular sample. The reports should examine actual or possible sources of error in the data collection procedures and the recording processes and their effects. These two reports could also provide evidence for the other aspects of data collection and recording which are listed in the following paragraphs.

Data collection sheets designed by the student which display the results of at least two data collection and measuring tasks:

- at least two tasks involving money
- at least two tasks involving measurement of physical dimensions
- at least two tasks involving data collection/measurement of one other property
- at least two further tasks, each involving measurement of a different rate of change (eg speed, inflation rates).

These reports should include evidence of extraction of data from at least two written sources and from at least two tasks involving collection of data from people.

Records or responses to oral questions put to the student which show the student's capacity to estimate quantities in relation to distance and size; at least two tasks to be completed for each of these.

Notes of working should show the student's capacity to work in a range of settings with a range of numbers (eg rounding to whole numbers; working to three decimal places), to understand and work with tolerances, and to use estimation to check results and predict outcomes.

Skills in this element should be learned and demonstrated through activities which will enhance the student's capacity to perform effectively in vocational settings.

AMPLIFICATION

Choose and use techniques (PC2): at this level, the student should be able to design and undertake successfully data collection procedures, deciding on appropriate levels of accuracy to work to (eg rounding to two significant figures in weighing a range of engineering components; working to three decimal places when working with data from measuring tasks). S/he should also be able to select an appropriate sample in order to obtain adequate, reliable data (eg how many people to ask in each category of a sample - for example, how many men under thirty, how many over thirty to ask about attitudes to driving; decide at what stages of production, and how frequently to sample items being put together on a production line).

Fractions (PC2 and PC3 range): examples include about $\frac{2}{3}$ (eg to describe the approximate proportion of customers spending over £50 in a shop), over $\frac{3}{4}$ (eg to describe the approximate proportion of telephone calls made from an office which are long distance and made during peak charge times).

Decimal fractions (PC2 and PC3 range): examples include 0.85m (eg when measuring the dimensions of an object), 0.35 l (eg when measuring out fluids).

Percentages (PC2 and PC3 range): examples include under 5% (eg to describe the approximate proportion of time a computer network is failing to work), over 80% (eg to describe the approximate proportion of people visiting a National Park during the winter period at weekends rather than weekdays).

Ratios (PC2 and PC3 range): examples include 2:5 (eg to describe two materials mixed together), 1:15 (eg to describe the average teacher-student ratio in teaching sessions in college departments).

Negative numbers (PC2 and PC3 range): examples include –0.5°C (eg when measuring daytime winter temperature), –£20,000 (eg when looking at end-of-year trading losses for a company).

Estimation to judge quantities and proportions (PC2 and PC3 range): the student should be able to estimate quantities and size when collecting data (eg 'this corridor is about 8.5 metres long'; 'if we stack up six of those boxes they will come to over 2m'; 'each file is just over 2cm thick'), and estimate fractions (eg 'that's about a quarter') and percentages (eg 'that's about 30%').

Estimation to check results (PC2 and PC3 range): this should be used by the student to check whether results are reasonable, using knowledge of the context and/or the size of numbers which might be expected.

Measuring instruments (PC2 and PC3 range): these should be selected by the student in order to collect the right sort of data at the level of accuracy which s/he has decided to be appropriate to the task (eg a thermometer which measures to a tenth of a degree; a steel rule with mm marked; weighing scales with grams marked).

Data collection procedure (PC2 and PC3 range): the emphasis is on designing valid and effective procedures for exploring problems and questions; the precise procedures can include using a questionnaire (eg of people's reactions to the level of service an organisation provides); a survey using observation (eg the routes people take through a building); or an experiment (eg how a material responds to changes in temperature).

From written sources (PC2 and PC3 range): examples include paper-based materials (eg a list of annual sales figures in a company's annual report; a table in a newspaper displaying the population of different towns) and IT-based materials (eg a chart from a database showing the temperature variation over a year; figures from a spreadsheet on annual profits).

From people (PC2 and PC3 range): examples include asking a sample of people how much they spend on specific products; asking people the distance they travel, the time they take and the amount it costs.

Discrete data, continuous data (PC2 and PC3 range): collecting and recording discrete data (eg numbers of children per couple), collecting and recording continuous data (eg measuring temperature changes in a room over time and deciding on appropriate intervals in order to create frequency tables).

Large data sets (PC2 and PC3 range): at this level, the student should be able to collect and work with large quantities of data which will require care and attention to manage effectively and accurately (eg extracting data from a large company's survey of more than 100 employees' experiences of training programmes run by a big local employer, examining a wide range of issues such as the nature of training, its duration, the outcomes, etc; working with data on the incidence of all classes of industrial injury in 20 examples of different types of employers in a locality, comparing it with national statistics).

Levels of accuracy (PC4): at this level, the student should undertake measuring tasks and the recording of data which involve nearest whole numbers (eg measuring the dimensions of a room to the nearest foot) and others which involve working to decimals (eg 0.4 1 when measuring liquid). At this level, the student should work with numbers of any size (eg population of a city 13,675,000; outside winter temperature of −3.4°C).

Tolerances (PC4 range): the student should be able to select and work to a suitable level of accuracy for data collection and measuring procedures which s/he has designed, and set and work within tolerances (eg ±5mm for a task measuring dimensions; ±15min timing an activity).

Units of physical dimensions (PC5 range): the student can work with either metric or imperial units. Examples include measuring width, depth and height of items, measuring distance between locations or objects. Note that in Element 3.2 the student will be expected to convert from one system to another.

Units of another property (PC5 range): examples include temperature (°F, °C), time (hours, minutes, seconds), metric units of capacity (l, ml), metric units of mass (kg, g), imperial units of mass (lb, oz).

Units of rate of change (PC5 range): examples include speed (eg m/sec), inflation (eg 6.5%), movement of materials (eg l/min when filling a pool with water).

Errors (PC7): the student should be aware of how errors can occur when information is collected: for example through inaccuracies in measuring procedures, the cumulative effects of using tolerances when measuring and recording; or through bias in the samples used. This level should give the student the opportunity to explore the effect of errors within large and small sets of data.

GUIDANCE

The tasks should be related to the settings in which students are operating. Examples of suitable opportunities include:

- surveying customers' opinions – through the selection of an appropriate sample and the design of a questionnaire – of existing food packaging and recording the results on a data collection sheet. Using estimating and measuring when designing new food packaging
- surveying the effect of information technology on the jobs of a large number of employees, selected from organisations of different sizes and sectors, recording the results on a data collection sheet.

■ Element 3.2: Tackle problems

The student decides which techniques to use to solve problems, including problems in three dimensions. In each case, solving the problem should involve using a series of techniques with numbers of any size, including negative numbers. The student uses both calculator and non-calculator methods to make calculations correctly and to check the results. The student decides which level of accuracy to use.

PERFORMANCE CRITERIA

A student must:

1 choose and use **techniques** which suit the problem
2 perform the **techniques** in a correct order
3 choose and use appropriate units
4 choose and use an appropriate **level of accuracy**
5 use mathematical terms correctly
6 carry out calculations correctly
7 use **checking procedures** to confirm the results of calculations
8 check that the results make sense in respect of the problem being tackled
9 identify the effects of any accumulating errors in calculations

RANGE

Techniques:

- **Number:** working with numbers of any size (using addition, using subtraction, using multiplication, using division); calculating (with fractions, with decimal fractions, with percentage, with ratios); calculations using simple formulas (expressed in words, expressed in symbols); using powers and roots
- **Shape, space and measures:** calculations using common units of measurement; solving problems in two dimensions (perimeters of shapes, the areas of plane shapes, solving problems in three dimensions (volume of simple shapes, volume of cylinders); calculations with compound measures
- **Handling data:** converting between common units of measurement using scales and tables, converting using calculations; calculating and using mean, median, mode and interquartile range

Levels of accuracy: selecting and using an appropriate level of accuracy; approximations

Units: approximate calculations, estimating, inverse operations

EVIDENCE INDICATORS

Work in the form of notes, sections of projects or reports, and/or workbooks, which shows detailed working in the procedures which the student used to tackle problems and tasks, and gives the reasons why particular approaches were chosen. The results of calculations should be accurate, and evidence of checking should be present; an evaluation of the effects of any accumulating errors in calculations should be included. Adequate evidence should be presented to show that the student is confident in the routine use of each technique listed in the range. This will mean the presentation of evidence which shows working with each technique in at least two tasks or problems.

The work should show that the student is able to work to different levels of accuracy in a range of settings, as demanded by the problems and tasks being tackled.

The skills in this element should be learned and demonstrated through activities which will enhance the student's capacity to perform effectively in vocational settings.

AMPLIFICATION

Choose and use techniques (PC1): the student should be able to make decisions about which technique to use in response to particular problems and tasks. At this level, the student should deal with problems which demand the use of a series of techniques (eg working out profit margins on a range of products made by a company, identifying those which are most susceptible to variation in the world market price of raw materials, those which are operating to very tight margins, and those which are enjoying market growth or reduction in demand). At this level, the student should be able to use all four operations ($+ - \times \div$) with numbers of any size.

Numbers of any size (PC1 and PC2 range): the student should be able to work with numbers of any size (eg 2,377,060; 2.79; 0.003); at this level, this should include understanding and using positive powers and standard form for very large numbers (eg 2.7×10^9m of cloth produced by a textile factory in a year; 7.4×10^{11} l of waste processed by a sewage plant in one month). Students should be able to work with positive and negative numbers (eg working with profit and loss accounts to establish whether or not a company is breaking even).

Fractions (PC1 and PC2 range): calculations with fractions (eg $\frac{9}{17}$; $\frac{7}{12}$). At this level, the student should be able to work effectively with fractions when solving problems (eg be aware that $\frac{9}{17}$ is smaller than $\frac{5}{6}$ when comparing quantities).

Decimal fractions (PC1 and PC2 range): calculations with decimal fractions (eg 0.11m, 0.36 l).

Percentages (PC1 and PC2 range): calculations with percentages (eg 72% of the cost of a product, 17.5% for the calculation of VAT and back-calculations of VAT) and calculation of proportional change (eg a comparison of birth and death rates shows that the population growth rate has changed from 2.7% in an area in 1985 to –1% in 1990).

Ratios (PC1 and PC2 range): calculating and comparing ratios (eg 1:22, 2:17 when working out and comparing teacher-student ratios in a college).

Formulas (PC1 and PC2 range): simple formulas expressed in words (eg net profit = sales – cost of goods sold – overhead expenses) and simple formulas expressed in symbols (eg $V = I \times R$ in electricity). At this level, the student should be competent at calculations arising from formulas such as finding u where $1200 = 456 + 8u$, arising from $C = s + 8u$.

Using powers and roots (PC2 range): working with problems where proportional increase or inverse relations are involved (eg working out which size box is cheaper to use when sending items, working out the internal volume of boxes in order to compare the number of items different boxes will hold and the cost of sending boxes of a particular size and weight).

Volume of simple shapes (PC1 and PC2 range): examples include calculating the volume of an L-shaped room in order to decide what size central-heating radiators should be used; calculating the volume of a rectangular tank and working out how much is in it when it is $\frac{1}{4}$ full.

Volume of cylinders (PC1 and PC2 range): the student should be able to calculate the volume of different cylinders and compare the variation in volumes (eg how the volume of a syringe 2cm in diameter and 8cm long compares with another which is also 8cm long but is 1cm in diameter).

Compound measures (PC1 and PC2 range): calculating and working with compound measures (eg speed [kph], customers per hour, words per square inch).

Converting (PC1 and PC2 range): this should be completed in two ways: within single systems (eg 1.5m = 1500mm; and 2ft = 24in) and across systems (eg 1ft = 0.3m, and 37°C = 98.6°F). At this level, students should be able to use conversion constants (eg 0.1m equals 10cm) and formulas (eg to convert from °Celsius to °Fahrenheit).

Interquartile range (PC1 and PC2 range): at this level students should be able to work with interquartile range in order to decide whether to include or exclude data in calculations (eg whether to include or exclude people in a sample who are odd extremes, for example, when they have unusually high income, which would distort the calculations on income in a workplace).

Levels of accuracy (PC4): at this level, the student is expected to select a suitable level of accuracy (eg rounding to whole numbers; working to three decimal places).

Approximations (PC4 range): examples include rounding to the nearest whole number, to the nearest 10.

Approximate calculations (PC7 range): for example, using whole number calculations which have been worked out using non-calculator methods in order to check whether the results of more precise calculations look correct.

Inverse operations (PC7 range): examples include checking subtraction by addition, and division by multiplication.

Accumulating errors (PC9): the student is expected to choose a level of accuracy appropriate to the task; when working with approximate numbers, or rounding to significant figures, the student should evaluate the scale of any accumulating errors, and their effect on the overall results of calculations.

GUIDANCE

The techniques could be performed on the data collected for Element 3.1 and should be related to the settings in which the student is operating. Examples include:

- calculating the volume of the packaging designed in Element 3.1 to ensure that it is sufficiently large. Calculating the area of cardboard to be used for the packaging design to contribute to the calculation of approximate costings. Calculating the percentages of people surveyed in Element 3.1 to ascertain the most popular colour, shape, material, method of opening and size and ranking these using inequality symbols
- grouping the data obtained from the large-scale survey of employees' opinions of information technology into organisation size, organisation sector, age of employee, level of job of employee. Calculating percentages of responses for each of the groups and for the whole data set and calculating the mean, median and mode responses for the whole set and for the individual groups.

■ Element 3.3 Interpret and present data

The student identifies the main features of the data and relates these to the problem tackled. The student decides from a wider range of more complex techniques which to use to display data and which titles and labels to use. The student decides what level of accuracy and which scales and axes to use.

PERFORMANCE CRITERIA

A student must:

1 identify and **explain the main features** of the data
2 choose and use **techniques** which will represent the data effectively
3 follow **conventions** for presenting the data
4 present the results with an appropriate **level of accuracy**
5 explain how the results make sense in respect of the problem being tackled

RANGE

Explain the main features: expressing in words the main relationships and patterns, expressing relationships in symbols, expressing relationships involving rate, expressing relationships as equations and inequalities

Techniques:

- **Number:** interpreting and presenting result of calculations; using probability to describe situations; interpreting and presenting upper and lower bounds of results
- **Shape, space and measures:** interpreting and constructing two-dimensional diagrams (plans, drawings, network diagrams); interpreting and constructing two-dimensional representations of three-dimensional objects
- **Handling data:** interpreting and constructing statistical diagrams (pictograms, bar charts, pie charts, histograms, scatter diagrams), interpreting and constructing graphs; interpreting and comparing mode, interpreting and comparing mean, interpreting and comparing median, interpreting and comparing range; working with interquartile range

Conventions: using given axes, selecting and using appropriate labels

Levels of accuracy: selecting and using appropriate units for presenting data; selecting and using appropriate scales

EVIDENCE INDICATORS

Written presentations of data, from written reports or spoken presentations, which display data using all of the different forms listed in the range for Shape, space and measures and Handling data. The presentation of data in this element can use the data collected in Element 3.1 and/or the results of calculations in Element 3.2.

Evidence should show that the student is confident in the use of each technique listed in the range. At least two examples of each form of diagram etc listed in Shape, space and measures and Handling data should be presented; these should demonstrate that the student is able to choose and use axes, scales and conventions appropriate to different settings.

For each piece of work, s/he should provide an explanation of the main features of the data and how these relate to the task or problem which is being tackled.

The skills in this element should be learned and demonstrated through activities which will enhance the student's capacity to perform effectively in vocational settings.

AMPLIFICATION

Main features (PC1): this relates to the key features of a set of data (eg trends – such as a rate of sales increasing over time; distribution – such as the way major industry is spread across a county; upper and lower limits – such as variation over a three-month period of temperature extremes in a building). It is essential that the student understands how the data relates to the problem being tackled (eg to indicate that while productivity will be increased by the introduction of some new equipment, the increase in productivity is not sufficient to offset the very high cost of purchase, installation, operation and maintenance). At this level, the student should express the relationships and patterns in words (eg 'when production of components in workshop team #1 reached the highest level of 1300 per week, the proportion of defective items increased dramatically, to over 12%; the same increase did not occur for workshop team #2'), in symbols (eg 'yield per acre = harvested crop divided by seed' expressed as $y = H/S$), expressing relationships involving rates (eg comparing growth rates in four different companies and dividends to share holders) and as equations (eg in pricing the cost of care for a patient in a hospital using the equation $P = T (M + H/N)$, where T = time spent in hospital by the patient in number of days, M = daily cost of the patient's medicine, H = total costs in running the hospital (heating, salaries, etc), and N = total number of patients in the hospital; using powers in showing the escalation in cost of reducing defects in products to 95% and above).

Data (PC1): Elements 3.1 and 3.2 will produce data which can be used as the basis for the tables, charts and other forms of presentation which are described in this element.

Choose and use techniques (PC2): the student should make decisions about which is the most appropriate technique for displaying data for specific tasks – see the range for Shape, space and measure and Handling data. The student should choose appropriate scales and axes for diagrams, and appropriate class intervals for grouped data. At this level, the student should also choose the level of accuracy which would be most appropriate for presenting data (eg round to whole numbers, display to two decimal places) in order to show the key features of the data effectively (eg to show that while the overall number of fatal car accidents has increased over a five-year period, the number of accidents per mile travelled has decreased, but the rate of decrease is slowing).

Using probability to describe situations (PC2 range): numerical descriptions of the probability of events (eg 3% probability to say that 3 items out of every 100 from a production line are defective) and use of the probability scale as a way of describing the probability of events (eg 0.5 as a way of saying that something will probably happen 10 times out of 20).

Upper and lower bounds (PC2 range): at this level, the student should be able to say where solutions or results lie (eg given the approximations used to plot projected profits for the coming year for a company, the result will lie between £120k and £95k).

Two-dimensional diagrams (PC2 range): the student should be able to construct plans (eg to show equipment layout in an office), and drawings (eg to show the most convenient and safe heights of storage areas for different chemicals).

Network diagrams (PC2 range): examples include a diagram showing most- and least-used routes to reading areas in a library, a diagram showing the shortest of three possible routes between two destinations.

Two-dimensional representations of three-dimensional objects (PC2 range): the student should be able to construct drawings which convey the dimensions of objects (eg a sketch drawing of a room to show the volume of the room and the amount of area taken by windows, in order to work out whether it is within current building regulations).

Statistical diagrams (PC2 range): diagrams which require the student to compare distributions and make inferences: pictograms (eg to show the balance of different types of industry in different parts of the country), bar charts (eg composite bar charts to show monthly production totals of four products over a year), and pie charts (eg to show incidence of respiratory illness in different age groups in a population), histograms (eg to show market share of 15 different companies), scatter diagrams (eg to show income of families against expenditure). At this level, the student should be interpreting and comparing data from different sources (eg comparing means from two graphs from different companies showing the salaries paid to different categories of workers).

Graphs (PC2 range): the student should be able to construct line graphs which allow comparisons of different categories to be drawn and compared on the same graph (eg to show the differing rates of use of paper in different departments over a three-month period, to show how the rate at which petrol consumption increases with speed differs for a range of small cars).

Working with interquartile range (PC2 range): deciding on which parts of a set of data to display in order to show the most representative picture for the task in hand (eg excluding certain extreme results when presenting survey results because they distort some calculations; including extreme results because they show the true range of results).

Axes (PC3 range): the student should choose axes in order to show data to the best effect, ensuring that independent variables (eg time) are displayed along the bottom (x) axis.

Labels (PC3 range): accurate and clear labels should be attached to all scales, axes and all other components of data.

Selecting and using appropriate scales (PC4 range): the student should be able to select scales which will allow important features of the data to be displayed effectively; this will include choosing scales of the appropriate intervals and units (eg deciding not to begin a scale at zero since the key data is temperature variation in a kiln between 900°C and 1200°C).

GUIDANCE

The data presented could be that collected in Element 3.1 and/or the results of the calculation of Element 3.2. Examples include:

- presenting the results of the survey into customer opinions about packaging using pictograms and bar charts to represent the percentages calculated in Element 3.2. Producing drawings of the packaging researched and designed in Element 3.1, accurately labelling the dimensions as measured in Element 3.1 and relating the colour, shape, material, method of opening and size to the results of the survey. Using graphs and statistical diagrams to represent the results from the large-scale survey of employees' opinions of information technology and to allow comparison of grouped data with the whole data set, suggesting reasons for any differences.

The following three tables show how the elements at both levels are covered in each part of the book.

Part 1: NUMBER

Element 1: Collect and record data

Level 2	Level 3
describing situations – using fractions – using decimal fractions – using percentages – using ratios – using negative numbers	describing situations – using fractions – using decimal fractions – using percentages – using ratios – using negative numbers
using estimation – to judge quantities – to judge proportions – to check results – to predict outcomes	using estimation – to judge quantities – to judge proportions – to check results – to predict outcomes

Element 2: Tackle problems

Level 2	Level 3
working with numbers of any size – using addition – using subtraction – using multiplication – using division	working with numbers of any size – using addition – using subtraction – using multiplication – using division
calculating – with fractions – with decimal fractions – with percentage – with ratios	calculating – with fractions – with decimal fractions – with percentage – with ratios
calculations using simple formulas – expressed in words – expressed in symbols	calculations using simple formulas – expressed in words – expressed in symbols
	using powers and roots

Element 3: Interpret and present data

Level 2	Level 3
interpreting and presenting result of calculations	interpreting and presenting result of calculations
using probability to describe situations	using probability to describe situations
	interpreting and presenting upper and lower bounds of results

Part 2: SHAPE, SPACE AND MEASURES

Element 1: Collect and record data

Level 2	Level 3
choosing and using appropriate measuring instruments and appropriate units of measurement for the task	choosing and using appropriate measuring instruments and appropriate units of measurement for the task

Element 2: Tackle problems

Level 2	Level 3
calculations using common units of measurement: solving problems in two dimensions – perimeters of shapes – the areas of plane shapes	calculations using common units of measurement: solving problems in two dimensions – perimeters of shapes – the areas of plane shapes
calculations using common units of measurement: solving problems in three dimensions – volume of simple shapes – volume of cylinders	calculations using common units of measurement: solving problems in three dimensions – volume of simple shapes – volume of cylinders
	calculations with compound measures

Element 3: Interpret and present data

Level 2	Level 3
interpreting and constructing two-dimensional diagrams – plans – drawings	interpreting and constructing two-dimensional diagrams – plans – drawings – network diagrams
interpreting and constructing two-dimensional representations of three-dimensional objects	interpreting and constructing two-dimensional presentations of three-dimensional objects

Part 3: HANDLING DATA

Element 1: Collect and record data

Level 2	Level 3
designing and using a data collection procedure for a given sample	selecting an appropriate sample and designing and using a data collection procedure
obtaining data from written sources obtaining data from people	obtaining data from written sources obtaining data from people
working with discrete and grouped data	discrete data continuous data
	handling large data sets

Element 2: Tackle problems

Level 2	Level 3
converting between common units of measurement – using scales – using tables	converting between common units of measurement – using scales – using tables
converting using calculations	converting using calculations
calculating and using – mean – mode – median – range	calculating and using – mean – mode – median – interquartile range

Element 3: Interpret and present data

Level 2	Level 3
interpreting and constructing statistical diagrams – pictograms – bar charts – pie charts	interpreting and constructing statistical diagrams – pictograms – histograms – bar charts – scatter diagrams – pie charts
interpreting and constructing graphs	interpreting and constructing graphs
interpreting mode interpreting mean interpreting median interpreting range	interpreting and comparing mode interpreting and comparing mean interpreting and comparing median interpreting and comparing range
	working with interquartile range
expressing in words the main relationships and patterns	expressing in words the main relationships and patterns
expressing relationships in symbols	expressing relationships in symbols
expressing relationships involving rate	expressing relationships involving rate
	expressing relationships as – equations – inequalities

Part 1
NUMBER

1 Estimation and approximation

Estimation to judge quantities and proportions

Estimation is something you will use very often in your chosen vocation. You may need to estimate the amount of room required for a filing cabinet or desk, the length of a car, the amount of paint needed to paint a sign, the amount of time left to complete a task.

Exercise 1.1

1 Estimate the lengths of the following items in centimetres and inches:

- (a) a desk
- (b) a book
- (c) a camera
- (d) a microwave oven
- (e) a tin of beans.

Now check your results by using a ruler or tape measure.

2 Estimate the lengths of the following items in metres and feet:

- (a) a car
- (b) a corridor
- (c) a room
- (d) a garden fence
- (e) a playing field.

Now check your results by using a ruler or tape measure.

3 Estimate the weights of the following items in grams and ounces:

(a) a personal stereo
(b) a book
(c) a child's toy
(d) a hammer
(e) a camera.

Now check your results by using a set of kitchen scales.

4 Estimate the weights of the following items in kilograms and pounds:

(a) a chair
(b) a vacuum cleaner
(c) a member of your family
(d) a garden spade
(e) a basket of laundry.

Now check your results by using a set of bathroom scales.

In your vocation you will often wish to estimate fractions or percentages of things. Sometimes it is not important to be accurate and an estimation will be sufficient.

EXAMPLE 1

5 out of 22 people visiting a newsagents do not purchase a newspaper.

The proprietor may say 'about a quarter of customers do not buy a newspaper'.

$\frac{5}{22}$ estimated is $\frac{5}{20} = \frac{1}{4}$

EXAMPLE 2

Out of 95 metal cases produced on a machine 30 are found to be faulty.

The quality control supervisor may say 'about 30% of the production run is faulty'.

$\frac{30}{95} \times \frac{100}{1} = \frac{3000}{95} = 31\frac{55}{95} = 31\frac{11}{19}\%$ estimated is $\frac{30}{90} \times \frac{100}{1} = \frac{3000}{90} = 30\%$

Exercise 1.2

Without using a calculator estimate the following as fractions.

1 A doctor sees 50 patients in one day. 12 of the patients required a prescription for antibiotics. About what fraction of patients required antibiotics?

2 A computer hardware engineer has 33 metres of cable to install in an office. He has a new roll of 300 metres of cable. About what fraction of the cable will he use?

3 Out of 37 jugs a potter makes, 4 are defective. About what fraction of jugs are defective?

4 A painter uses 2.3 litres of paint out of a 5 litre tin. About what fraction of paint has been used?

5 An entrepreneur has £39 000 start up capital. She spends £12 750 on machinery. About what fraction of the start up capital is being spent on machinery?

Without using a calculator estimate the following as percentages.

6 In a week a nurse spends 9 hours in the casualty department and 31 hours in the radiology department. About what percentage of her time is spent in casualty?

7 A photographer takes 33 wedding photographs. 8 of the photographs are of the bride only. About what percentage of the photographs are of the bride only?

8 A baker bakes 250 loaves of bread. 105 are white, 70 are wholemeal and 75 are granary. About what percentage of the 250 bread loaves is white bread?

9 Passengers on an aeroplane are allowed one suitcase per passenger. 12 out of 125 passengers have taken more than the baggage allowance. About what percentage of passengers have taken more than the baggage allowance?

10 A builder receives a delivery of 20 000 bricks. 425 of the bricks are found to be broken upon delivery. About what percentage of the bricks are broken?

Estimation to check results and predict outcomes

You should use this to check whether results are reasonable, using knowledge of the context and/or the size of numbers which might be expected.

EXAMPLE 3

A personnel officer gave a questionnaire about weekly working patterns to five employees. Her calculation of the mean hours worked was 51 hours. She thought this to be too high as the minimum number of hours each employee has to work is 40. She went back over the data to check the result and discovered that one of the employee's working hours were very high.

Employee	Hours worked per week
1	44
2	45
3	48
4	72
5	46

After talking to Employee 4 she discovered that he did not take the questionnaire seriously and that he had exaggerated his answers.

Exercise 1.3

Without using a calculator say whether the following results are reasonable.

1 A personal trainer makes bookings with clients for 1-hour sessions. On one particular day he has 11 bookings. Is this reasonable? Why?

2 A mail order company receives, on average, five orders per day for squeegee mops. The warehouse manager suggests they keep a stock of 1000 and reorder every three months. Is this reasonable? Why?

3 A secretary can type 80 words per minute. Her employer wishes her to type a document which contains 10 000 words. He wants it within three hours. Is this reasonable? Why?

4 A newly appointed local authority housing officer discovers that 3 800 council tenants are in arrears with their rent. He determines to reduce this number by half within 6 months. Past statistics are:

Year	Quarter	Number in arrears
19–1	1	5000
	2	4750
	3	4740
	4	3500
19–2	1	3700
	2	3950
	3	3900
	4	3800

Is this a reasonable determination? Why?

5 Advertising literature for a laser printer states that it will print 6–9 pages per minute. A secretary estimates that it will take 5 minutes to print a 40 page document. Is this reasonable? Why?

Approximations and approximate calculations

It is important that you get into the habit of approximating your answers when using a calculator. It is very easy to press a wrong key on a calculator thus getting an incorrect answer. However, if you have approximated you will know the answer is wrong.

EXAMPLE 4

Rounding to the nearest whole number:

- 1.1 m becomes 1 m
- £25.27 becomes £25
- £56.85 becomes £57
- 7.8 ml becomes 8 ml.

Rounding to the nearest 10:

- 87 becomes 90
- £123.30 becomes £120
- 12 784 becomes 12 780.

Rounding to the nearest 100:

- 12 784 becomes 12 800
- 589 becomes 600
- 715 becomes 700.

Exercise 1.4

Approximate to the nearest whole number:

1 £3.27

2 5.75 m

3 33.9 °C

4 24.67 l

5 489.23 km.

Approximate to the nearest 10:

6 £143 .57

7 382 m

8 173 kg/m^2

9 1274 km

10 585 books.

Approximate to the nearest 100:

11 27 482 miles

12 £135 229

13 13 744 bricks

14 8551 leaflets

15 62 587 kb of disk space free.

EXAMPLE 5

Three items bought from a camping store totalled £585.29:

Tent £375.85; Gortex jacket £167.99; Walking boots £41.45.

Approximated total is £584:

Tent £375: Gortex jacket £168; Walking boots £41.

Exercise 1.5

Without using a calculator approximate the following:

1 3.79 m + 2.12 m

2 £29.57 + £48.22

3 75.21 kg – 4.98 kg

4 78 432 kb of disc space – 5389 kb

5 £149 999 + £150 123 + £79 898

■ Inverse operations

You are expected to be able to check any calculation with the inverse operation.

For example: checking subtraction by addition (− by +)

checking addition by subtraction (+ by −)

checking division by multiplication (÷ by ×)

checking multiplication by division (× by ÷)

■ Upper and lower bounds

Instead of giving one approximate figure advanced level students are expected to give an upper approximation and a lower approximation. You are expected to be able to say where solutions or results lie.

EXAMPLE 6

A house builder requires 19 822 bricks. Three bricks have been identified as possible choices.

Brick A costs 43p, brick B costs 22p and brick C costs 37p.

Approximate the 19 822 bricks to 20 000.

Brick A   43p × 20 000 = 860 000p

$$= £8600$$

Brick B   22p × 20 000 = 440 000p

$$= £4400$$

The bricks will approximately cost between £4400 and £8600.

Exercise 1.6

1 A machine makes 4800 car parts in any one day. The cost of each part depends upon the number purchased. Each part is sold for either £2.20, £2, £1.75 or £1.45. Calculate the approximate upper and lower revenue expected in any one day.

2 Four employees' approximate weekly wages are £379, £405, £298 and £318. In some weeks they each earn £10 less than this figure and in other weeks they earn £20 more. Calculate the approximate upper and lower wages bills the company may expect.

3 A pharmaceutical company proposes to launch a new medicine. They wish to sell the medicine for either £31 a litre or £39 a litre. How many litres will they have to sell to achieve a revenue of £960 000?

2 Working with numbers of any size

Addition and subtraction

After counting, addition and subtraction were the first mathematical techniques you learned as a young child. You will continue to use these techniques throughout your working life; therefore it is important that you can add and subtract both single and double digit numbers mentally. For example, if you have difficulty adding 10 + 35 + 44 mentally you should start practising today. You should also be able to add slightly more difficult numbers without a calculator, e.g. using pencil and paper methods.

In the workplace it is not always possible to put your hands immediately on a calculator, therefore it is important that you can carry out calculations on paper or in your head. Even when you do have access to a calculator it is also important to be able to add and subtract mentally to check your answer is correct (you may press the wrong key, for instance).

Exercise 2.1

1 Mentally calculate the following:

(a) 5 + 7 + 9 =	(f) 25 − 12 =	(k) 17 − 9 =
(b) 14 + 3 + 7 =	(g) 40 − 14 =	(l) 35 − 23 =
(c) 9 + 21 + 2 =	(h) 29 − 17 =	(m) 65 − 29 =
(d) 4 + 7 + 12 + 8 + 16 + 3 =	(i) 49 − 15 =	(n) 24 − 15 =
(e) 6 + 18 + 24 + 35 + 7 =	(j) 14 − 6 =	(o) 16 − 7 =

2 A manufacturing company buys three items of office furniture: a desk costing £422, a chair costing £216 and a filing cabinet costing £462. How much did they spend altogether?

3 A nursing home buys new linen to replace items that have become old and worn out. They buy 144 sheets, 96 duvet covers, 72 pillow cases, 24 tablecloths and 12 teatowels. How many items did they buy?

4 The receipts for a car service centre in one day's trading are shown in Table 2.1.

Table 2.1

Name	£	Method of payment
Mrs Norris	189.60	Cash
Mr Ediberi	220.89	Cheque
The Ice-Cream Centre	380.00	Account
Mr Igwe	55.00	Credit Card
Miss Thompson	20.00	Cash
Mrs Evans	527.31	Credit Card
Mr Thaper	156.80	Cheque
Fox & Co.	150.00	Account
Mr Schultz	82.67	Cash
Mrs Grant	189.60	Credit Card
Miss Holland	95.48	Cash
The Garden Centre	135.00	Account
Mrs Johnson	160.25	Cheque
Mr Stewart	672.87	Credit Card

What were the following totals:

(a) Cash receipts
(b) Cheque receipts
(c) Credit Card receipts
(d) Account receipts
(e) Total receipts?

5 The members of a fitness club held a sponsored step. Add up the number of steps achieved by each participant to get the total number for the event.

Pat Chang	21 360
John Hargreeves	31 580
Amy Davies	11 230
Jessica Long	31 445
Jon Yoko	21 400
Sam Wright	15 892

6 A manufacturer of kites employs 5 machinists to sew fabric stunt kites. From the weekly worksheet shown in Table 2.2 find:

(a) the machinist making the greatest number of kites
(b) the total number of kites made.

Table 2.2

	Mon	Tues	Wed	Thur	Fri
Machinist A	65	67	66	68	72
Machinist B	73	72	70	55	ill
Machinist C	60	69	70	75	65
Machinist D	71	71	71	71	71
Machinist E	58	58	59	57	60

7 During one day of trading an electrical superstore had the opening and closing stock levels shown in Table 2.3.

What quantity of each item was sold that day?

8 An upholsterer had 500 metres of brown fabric. He used 27 m to cover a three piece suite, 4.5 m to cover 4 dining room chairs, 8.5 m to cover a fireside chair and 16.25 m to cover a set of caravan cushions. How much fabric was remaining?

9 A silversmith applies three layers of silver plate to a candlestick. The thickness of the three layers are 0.073 mm, 0.051 mm and 0.003 mm respectively. What is the total thickness of the silver plate?

Table 2.3

Item	Opening stock	Closing stock
Portable cassette player	49	37
Portable CD player	28	9
Mini hi-fi system	15	7
Midi hi-fi system	12	3
Car radio	13	9
Television	24	17
Video player	32	24
Satellite system	11	8
Washing machine	21	15
Tumble dryer	17	3
Refrigerator	18	12
Freezer	11	7
Vacuum cleaner	36	19

10 A warehouse has 4 large storage boxes full of electrical components. The number of components each storage box contains is 2 478 328, 4 784 391, 1 785 899 and 3 472 673 respectively. What is the total number of components stored at the warehouse?

Multiplication and division

Along with addition and subtraction, you should be able to calculate multiplication and division mentally . This involves having instant recall of the 1 to 12 times tables and a working method of calculating higher tables. If you cannot recall all the 1 to 12 times tables then start practising now with the help of the tables reproduced for you on page 16.

1 x table					2 x table					3 x table					4 x table			
1 x 1 =	1		1 x 2 =	2		1 x 3 =	3		1 x 4 =	4
2 x 1 =	2		2 x 2 =	4		2 x 3 =	6		2 x 4 =	8
3 x 1 =	3		3 x 2 =	6		3 x 3 =	9		3 x 4 =	12
4 x 1 =	4		4 x 2 =	8		4 x 3 =	12		4 x 4 =	16
5 x 1 =	5		5 x 2 =	10		5 x 3 =	15		5 x 4 =	20
6 x 1 =	6		6 x 2 =	12		6 x 3 =	18		6 x 4 =	24
7 x 1 =	7		7 x 2 =	14		7 x 3 =	21		7 x 4 =	28
8 x 1 =	8		8 x 2 =	16		8 x 3 =	24		8 x 4 =	32
9 x 1 =	9		9 x 2 =	18		9 x 3 =	27		9 x 4 =	36
10 x 1 =	10		10 x 2 =	20		10 x 3 =	30		10 x 4 =	40
11 x 1 =	11		11 x 2 =	22		11 x 3 =	33		11 x 4 =	44
12 x 1 =	12		12 x 2 =	24		12 x 3 =	36		12 x 4 =	48

5 x table					6 x table					7 x table					8 x table			
1 x 5 =	5		1 x 6 =	6		1 x 7 =	7		1 x 8 =	8
2 x 5 =	10		2 x 6 =	12		2 x 7 =	14		2 x 8 =	16
3 x 5 =	15		3 x 6 =	18		3 x 7 =	21		3 x 8 =	24
4 x 5 =	20		4 x 6 =	24		4 x 7 =	28		4 x 8 =	32
5 x 5 =	25		5 x 6 =	30		5 x 7 =	35		5 x 8 =	40
6 x 5 =	30		6 x 6 =	36		6 x 7 =	42		6 x 8 =	48
7 x 5 =	35		7 x 6 =	42		7 x 7 =	49		7 x 8 =	56
8 x 5 =	40		8 x 6 =	48		8 x 7 =	56		8 x 8 =	64
9 x 5 =	45		9 x 6 =	54		9 x 7 =	63		9 x 8 =	72
10 x 5 =	50		10 x 6 =	60		10 x 7 =	70		10 x 8 =	80
11 x 5 =	55		11 x 6 =	66		11 x 7 =	77		11 x 8 =	88
12 x 5 =	60		12 x 6 =	72		12 x 7 =	84		12 x 8 =	96

9 x table					10 x table					11 x table					12 x table			
1 x 9 =	9		1 x 10 =	10		1 x 11 =	11		1 x 12 =	12
2 x 9 =	18		2 x 10 =	20		2 x 11 =	22		2 x 12 =	24
3 x 9 =	27		3 x 10 =	30		3 x 11 =	33		3 x 12 =	36
4 x 9 =	36		4 x 10 =	40		4 x 11 =	44		4 x 12 =	48
5 x 9 =	45		5 x 10 =	50		5 x 11 =	55		5 x 12 =	60
6 x 9 =	54		6 x 10 =	60		6 x 11 =	66		6 x 12 =	72
7 x 9 =	63		7 x 10 =	70		7 x 11 =	77		7 x 12 =	84
8 x 9 =	72		8 x 10 =	80		8 x 11 =	88		8 x 12 =	96
9 x 9 =	81		9 x 10 =	90		9 x 11 =	99		9 x 12 =	108
10 x 9 =	90		10 x 10 =	100		10 x 11 =	110		10 x 12 =	120
11 x 9 =	99		11 x 10 =	110		11 x 11 =	121		11 x 12 =	132
12 x 9 =	108		12 x 10 =	120		12 x 11 =	132		12 x 12 =	144

Exercise 2.2

1 Mentally calculate the following:

(a) $6 \times 9 =$ (f) $8 \times 6 =$ (k) $36 \div 9 =$

(b) $3 \times 8 =$ (g) $9 \times 7 =$ (l) $24 \div 6 =$

(c) $4 \times 7 =$ (h) $4 \times 9 =$ (m) $21 \div 7 =$

(d) $9 \times 11 =$ (i) $7 \times 6 =$ (n) $220 \div 20 =$

(e) $12 \times 7 =$ (j) $9 \times 8 =$ (o) $49 \div 7 =$

2 The stock from The Bed Warehouse was completely damaged by floods, and had to be destroyed. From the records of items, quantity and selling price shown in Table 2.4, calculate how much the shop claimed from the insurers.

Table 2.4

Item	Quantity	Selling price (£)
2′ 6″ Divan	12	82.50
2′ 6″ Bunkbed	9	150.79
3′ Divan	35	109.20
3′ Bunkbed	27	160.85
Double divan	47	205.39
Double four poster	2	419.27
King size divan	19	375.80

3 A plumber needs to lay 27 m of 22 mm thick copper pipe and 132 m of 15 mm thick copper pipe for a central heating system. A length of pipe is 3 m long and costs £5.69 for 22 mm and £3.45 for 15 mm.

Calculate:

(a) how many lengths of 22 mm pipe are required

(b) how many lengths of 15 mm pipe are required

(c) the total cost.

4 A builder wishes to build a wall 6.16 m long by 1.04 m high. If a brick measures 22 cm by 6.5 cm, calculate the following:

(a) How many bricks are required?

(b) If a brick costs 32p what is the cost of the bricks?

5 A theatre seats 1500 people. There are 250 box seats at £28.75 each, 450 stall seats at £22.25 each and the remainder in the balcony at £18.50 each. If the theatre has eight shows in a week with all performances sold out, what would be the total revenue for that week?

Negative numbers

A negative number is a number which is less than zero. For example, when measuring temperature in the winter the thermometer may read – 0.5 °C, or when looking at end-of-year trading figures for a company they may have made a loss of – £20 000.

Study the number line shown in Fig 2.1.

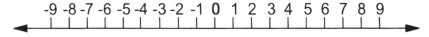

Fig 2.1

To add a number you travel *right* along the number line and to subtract a number you travel *left*.

EXAMPLE 1

3 + 4 = 7

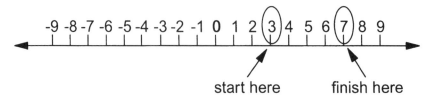

start here   finish here

Start at 3 and travel right 4 places.

$5 - 3 = 2$

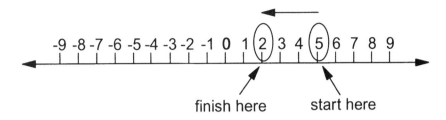

Start at 5 and travel left 3 places.

$3 - 8 = -5$

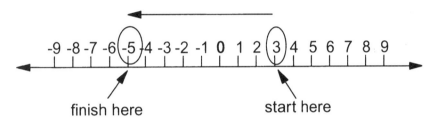

Start at 3 and travel left 8 places.

$-2 - 4 = -6$

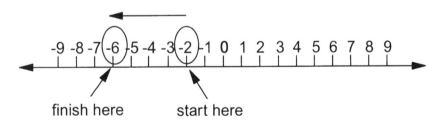

Start at -2 and travel left 4 places.

Obviously it is not easy to use a number line with very large numbers. Many calculators have a facility to allow you to put in negative numbers.

EXAMPLE 2

A company's records showed profit/loss over a three year period as being:

Year 1 – £358 492 (loss)

Year 2 – £349 286 (loss)

Year 3 – £563 739 (profit)

The +/– button on a calculator will allow you to input the negative numbers.

Enter 358492 and press the +/– button

+

349286 and press the +/– button again

+

563739

Answer = – £144 039

Exercise 2.3

1 Calculate: – 37 – 19 + 41 + 23 – 109.

2 Calculate: – 56 482 – 64 892 + 43 896 – 83 942 + 78 269.

3 The monthly balance between income and expenditure in a nursing home is shown:

Month	Balance
January	– £585
February	– £457
March	– £129
April	£35
May	£67
June	£243

What is the total balance at the end of the six months?

4 The balance between bets lost and bets won at a turf accountants are shown:

Day of week	Balance
Monday	– £38
Tuesday	£2859
Wednesday	– £1543
Thursday	£4776
Friday	£2649
Saturday	– £2482

What is the total balance at the end of the week?

5 A blood bank is constantly using and restocking its supply of bags of blood. The use and supply of O+ blood over a short period of time is:
– 32 bags; – 58 bags; + 72 bags; – 68 bags; + 134 bags; + 89 bags;
– 27 bags; + 46 bags; + 320 bags; – 16 bags.

How many bags of O+ blood are there in stock?

Standard form

Organisations which deal with very large numbers or quantities may wish to show these figures in standard form. For example, a textile factory which produces 2 700 000 000 metres of cloth a year may wish to show this as 2.7×10^9 m. A sewage plant may process 740 000 000 000 litres of waste in one month and they may wish to show this as 7.4×10^{11} litres.

To express a large number in standard form you must start with a value between 1 and 10. This value is then multiplied by 10 to a power.

EXAMPLE 3

Express 5 000 000 in standard form.

5 is a value
between 1 and 10

6 zeros give us 10^6

Answer = 5×10^6

Remember

$10^2 = 10 \times 10 = 100$

$10^3 = 10 \times 10 \times 10 = 1000$

$10^4 = 10 \times 10 \times 10 \times 10 = 10\ 000$

and so forth.

EXAMPLE 4

Express 482 231 000 in standard form.

4.8 is a value
between 1 and 10

4.82231000

decimal point has moved 8 places
left, therefore 10^8

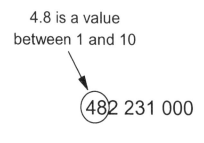

Answer = 4.8×10^8

EXAMPLE 5

Express 6 998 362 in standard form.

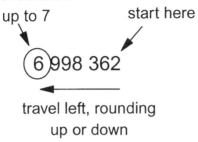

6 is rounded up to 7

start here

6 998 362

travel left, rounding up or down

Answer = 7×10^6

Exercise 2.4

1 Express in standard form:

    (a) 800 000 000
    (b) 27 000 000
    (c) 4 900 000
    (d) 123 000 000 000
    (e) 7 987 975 342 748.

2 Express as 'normal' numbers:

    (a) 4×10^5
    (b) 3.2×10^7
    (c) 9.1×10^{11}
    (d) 7.9×10^9
    (e) 1.1×10^{15}.

3 The speed of light is 299 792 500 m/s. Express this in standard form.

4 The population of China is 1 120 000 000. Express this in standard form.

5 The distance between the earth and the sun is 149 597 870 000 metres. Express this in standard form.

3 Fractions, decimal fractions, percentages and ratios

Fractions

A fraction is a small piece or any part of a whole. It has two main parts: a numerator and a denominator. You can remember it by the acronym NOD.

Numerator

Over

Denominator

The *denominator* tells you how many parts the whole has been divided up into and the *numerator* tells you how many parts of the whole are actually used. For example, a numerator 3 and a denominator 8 will give us $\frac{3}{8}$ as represented in Fig 3.1.

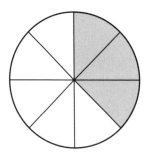

Fig 3.1 Representation of $\frac{3}{8}$

Types of fractions

There are three types of fractions: a proper fraction, an improper fraction and a mixed number.

- A *proper fraction* has the numerator smaller than the denominator and therefore is always less than 1: e.g. $\frac{3}{8}$.
- An *improper fraction* has the numerator larger than the denominator and is therefore always greater than 1: e.g. $\frac{12}{5}$.
- A *mixed number* is a mixture of whole numbers and a proper fraction and is therefore always greater than 1: e.g. $2\frac{2}{5}$.

An improper fraction can be changed to a mixed number and a mixed number can be changed to an improper fraction, e.g. $\frac{12}{5}$ is the same as $2\frac{2}{5}$.

EXAMPLE 1

(a) Convert $2\frac{3}{4}$ into an improper fraction.

The denominator tells you a whole number is to be divided into 4 parts. Two whole numbers with 4 parts in each equals 8 parts $(2 \times 4 = 8)$. Add the 3 parts from the fraction to the 8 parts from the whole $(3 + 8 = 11)$.

Answer $= \frac{11}{4}$

(b) Convert $\frac{14}{3}$ into a mixed number.

Divide the numerator by the denominator to find how many whole numbers there are $(14 \div 3 = 4, \text{ remainder } 2)$

Answer $= 4\frac{2}{3}$.

Fractions in their lowest form

It is always best to display fractions in their lowest form: e.g. $\frac{4}{8}$ becomes $\frac{1}{2}$ in its lowest form. You must divide both numerator and denominator by the same number. The most common numbers to look for are multiples of 2, 3, 5 or 10.

EXAMPLE 2

Cancel $\frac{16}{48}$ down into its lowest form.

Divide both the numerator and the denominator by 16 ($16 \div 16 = 1$, $48 \div 16 = 3$).

Answer $= \frac{1}{3}$

■ Comparing fractions

In order to compare two or more fractions you need to be aware of the sizes of the numerators and the denominators.

EXAMPLE 3

(a) If the fractions have the same size denominator then the largest numerator is the larger fraction, e.g.

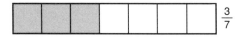

Fig 3.2 $\frac{3}{7}$ **and** $\frac{5}{7}$

Answer $= \frac{5}{7}$ is the larger fraction

(b) If the fractions have the same size numerator then the smallest denominator is the larger fraction, e.g.

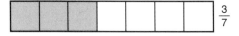

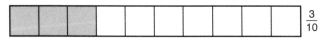

Fig 3.3 $\frac{3}{7}$ **and** $\frac{3}{10}$

Answer $= \frac{3}{7}$ is the larger fraction

Use the addition of fractions method to help you compare two or more fraction sizes.

■ Addition and subtraction of fractions

You need to be able to add and subtract fractions; this should preferably be without the use of a fraction button on a calculator.

- When adding or subtracting fractions that have the same denominator you simply add or subtract the numerators.
- When adding or subtracting fractions with different denominators you have to find a lowest common denominator (LCD). This is the lowest number into which both denominators will divide. Whatever number you multiply the denominator by to get the LCD, you must multiply the numerator by the same number.

EXAMPLE 4

(a) $\dfrac{3}{8} + \dfrac{2}{8} = \dfrac{5}{8}$

(b) $\dfrac{1}{3} + \dfrac{2}{5}$

15 is the lowest number that both 3 and 5 will divide into.

$$\frac{1}{3} + \frac{2}{5} = \frac{(1 \times 5)}{(3 \times 5)} + \frac{(2 \times 3)}{(5 \times 3)} = \frac{5}{15} + \frac{6}{15} = \frac{11}{15}$$

(c) $\dfrac{2}{3} - \dfrac{1}{6} = \dfrac{4}{6} - \dfrac{1}{6} = \dfrac{3}{6} = \dfrac{1}{2}$

6 is the LCD; it can be divided by itself and also by 3.

■ Multiplication and division of fractions

You need to be able to multiply and divide fractions.

- To multiply fractions you multiply the numerators together and multiply the denominators together.
- To divide fractions, you have to invert (turn upside down) the right-hand fraction and then multiply the fractions together.

EXAMPLE 5

(a) $\dfrac{2}{3} \times \dfrac{4}{5} = \dfrac{8}{15}$

(b) $\dfrac{2}{7} \div \dfrac{1}{3} = \dfrac{2}{7} \times \dfrac{3}{1} = \dfrac{6}{7}$

When multiplying or dividing mixed fractions, it is always advisable to convert the fractions to improper fractions. This saves making unnecessary mistakes.

Exercise 3 .1

1 Convert the following mixed numbers into improper fractions:

(a) $3\frac{1}{2}$ (b) $4\frac{3}{5}$ (c) $5\frac{1}{4}$ (d) $2\frac{8}{11}$ (e) $3\frac{4}{7}$

2 Convert the following improper fractions into mixed numbers:

(a) $\frac{11}{4}$ (b) $\frac{22}{3}$ (c) $\frac{19}{8}$ (d) $\frac{17}{5}$ (e) $\frac{23}{6}$

3 Cancel down the following into their lowest form:

(a) $\frac{8}{32}$ (b) $\frac{24}{48}$ (c) $\frac{28}{36}$ (d) $\frac{77}{112}$ (e) $\frac{102}{132}$

4 Which is the larger of the following sets of fractions?

(a) $\frac{4}{5}$ and $\frac{1}{5}$ (b) $\frac{2}{3}$ and $\frac{2}{7}$ (c) $\frac{1}{8}$ and $\frac{4}{5}$ (d) $\frac{7}{9}$ and $\frac{5}{11}$ (e) $\frac{6}{13}$ and $\frac{11}{25}$

5 Add the following fractions:

(a) $\frac{2}{3} + \frac{1}{4}$ (b) $\frac{1}{2} + \frac{5}{6}$ (c) $\frac{3}{4} + \frac{4}{5}$ (d) $\frac{1}{3} + \frac{2}{5} + \frac{1}{2}$ (e) $\frac{1}{4} + \frac{1}{6} + \frac{2}{3}$

6 Subtract the following fractions:

(a) $\frac{3}{8} - \frac{1}{4}$ (b) $\frac{11}{16} - \frac{5}{8}$ (c) $\frac{3}{5} - \frac{1}{7}$ (d) $\frac{5}{6} - \frac{4}{9}$ (e) $\frac{5}{8} - \frac{7}{16}$

7 Multiply the following fractions:

(a) $\frac{1}{2} \times \frac{4}{13}$ (b) $\frac{4}{7} \times \frac{4}{9}$ (c) $\frac{12}{17} \times \frac{2}{3}$ (d) $\frac{9}{20} \times \frac{25}{27}$ (e) $\frac{17}{22} \times \frac{11}{34}$

8 Divide the following fractions:

(a) $\frac{3}{8} \div \frac{5}{8}$ (b) $\frac{7}{16} \div \frac{1}{4}$ (c) $\frac{13}{21} \div \frac{3}{7}$ (d) $\frac{15}{18} \div \frac{13}{12}$ (e) $\frac{4}{9} \div \frac{24}{27}$

Decimal fractions

Whole numbers and decimals are constructed as shown in Fig 3.4.

Decimal point ↓

1,000's	100's	10's	Units	•	$\frac{1}{10}$	$\frac{1}{100}$	$\frac{1}{1000}$

Fig 3.4

If we put 0.235 into the table, we obtain the result shown in Fig 3.5.

1,000's	100's	10's	Units	•	$\frac{1}{10}$	$\frac{1}{100}$	$\frac{1}{1000}$
			0	•	**2**	**3**	**5**

Fig 3.5

$$0 \text{ units plus } \frac{2}{10} \text{ plus } \frac{3}{100} \text{ plus } \frac{5}{1000}$$

or

$$\frac{235}{1000}$$

As you can see a decimal is a fraction, but it is written without a denominator.

■ Adding and subtracting decimals

The golden rule to remember when adding and subtracting decimals on paper is to *always* line up the decimal points and numbers in columns.

EXAMPLE 6

(a) 0.25 + 2.375 + 73.425 + 137.5

$$
\begin{array}{r}
0.25 \\
2.375 \\
73.425 \\
137.5 \\
\hline
213.550 \\
\hline
\end{array}
$$

(b) 967.825 − 63.25

$$
\begin{array}{r}
967.825 \\
63.25 \\
\hline
904.575 \\
\hline
\end{array}
$$

If you use graph paper when adding and subtracting decimals, the squares will help you get all the numbers in the correct columns.

■ Multiplying decimals

When multiplying two decimal numbers:

1   Count the total number of decimal places in the numbers being multiplied.
2   Disregard the decimal points when multiplying the two numbers.
3   Replace the decimal point in the product by counting the places from right to left.

EXAMPLE 7

Multiply 3.25 by 5.2.

Total number of decimal places is 3. Ignore the decimal points.

$$\begin{array}{r} 325 \\ 52 \\ \hline 16900 \end{array}$$

Replace the decimal point (3 places from right).

Answer = 16.900

■ Dividing decimals

When dividing one decimal number by another:

1 Make the divisor into a whole number. Do this by moving the decimal point from left to right.
2 What you do to one number, you must do to the other. Move the decimal point in the number being divided, the same number of places from left to right.
3 Divide in the usual way.

EXAMPLE 8

Divide 4.505 by 0.85.

0.85 becomes 85   (decimal point moves two places to right)

4.505 becomes 450.5   (this decimal point must also move two places to the right)

$$\begin{array}{r} 5.3 \\ 85 \overline{\smash{)}\ 450.5} \end{array}$$

Answer = 5.3

Exercise 3.2

Do not use a calculator.

1 Add the following decimals:

(a) 0.969 + 0.661

(b) 208.16 + 23.719

(c) 5.74 + 39.3 + 2.63

(d) 4374.46 + 25.2 + 675.06

(e) 0.434 + 64.8 + 744.2 + 10.3

2 Subtract the following decimals:

(a) 457.87 − 80.07

(b) 27.4 − 6.3

(c) 731.649 − 210.23

(d) 50 889.51 − 8424.11

(e) 78 189.519 − 38 836.017

3 Multiply the following decimals:

(a) 47.6×0.3

(b) 361.3×7.2

(c) 0.518×34.6

(d) 0.048×0.178

(e) 2919.96×8.35

4 Divide the following decimals:

(a) 0.924 by 0.7

(b) 17.6 by 0.02

(c) 696.6 by 12.9

(d) 142.272 by 31.2

(e) 13.9854 by 0.22

Percentages

A percentage is a fraction with a denominator of 100: e.g. 33% is $\frac{33}{100}$.
It is a convention to use the % sign instead of the denominator.

- To change a percentage into a fraction or decimal, divide it by 100.
- To change a fraction or decimal into a percentage, multiply it by 100.

EXAMPLE 9

(a) Express 75% as a fraction.

Answer = $\frac{75}{100}$

This will cancel down to $\frac{3}{4}$.

(b) Express 6% as a decimal.

$6 \div 100$

Answer = 0.06

(c) Express $\frac{8}{25}$ as a percentage.

$\frac{8}{25} \times 100 = 32$

Answer = 32%

(d) Express 0.075 as a percentage.

0.075×100

Answer = $7\frac{1}{2}\%$

To find the percentage of a quantity you must first convert the percentage into a fraction or a decimal number. When you have done that you multiply by the quantity.

EXAMPLE 10

(a) Calculate 30% of 80 grams.

Convert to a decimal number.

$30\% = 30 \div 100 = 0.3$

$0.3 \times 80 = 24$

Answer = 24 grams

(b) Calculate 15% of £220.

Convert to a fraction.

$$15\% = \frac{15}{100} = \frac{3}{20}$$

$$\frac{3}{20} \times 220 = \frac{660}{20} = 33$$

Answer = £33

(c) A football team won 23 games out of 40. What percentage is this?

23 out of 40 is a fraction.

$$\frac{23}{40} \times 100 = 57\frac{1}{2}\%$$

Answer = $57\frac{1}{2}\%$

(d) 20% of a length of rope measures 600 cm. What is the complete length?

20% as a decimal = 0.2

Let the complete length be x. Therefore 20% of x = 600 cm.

$$0.2 \times x = 600$$

$$x = 600 \div 0.2 = 3000$$

Answer = 3000 cm

 Exercise 3.3

1 Calculate:

(a) 5% of 5420

(b) 20% of 380

(c) 80% of 400 g

(d) 67% of 13 500 m

(e) 42% of £415

2 Calculate:

    (a) 105% of 520   (d) 6% of 1500 km

    (b) 220% of 30   (e) $\frac{1}{2}$% of £120

    (c) 180% of 400 kg

3 Calculate the percentage for:

    (a) 25 out of 125   (d) 568 g out of 1704 g

    (b) 5 out of 20   (e) 356 people out of 800 people

    (c) 45 out of 375

4 Calculate the value for x if:

    (a) 25% of x is 250   (d) $12\frac{1}{2}$% of x km is 5948 km

    (b) 55% of x is 4125   (e) 66.7% of x ml is 6003 ml

    (c) 47% of £x is £40.42

Ratios

A ratio is a proportion. Cement mortar is mixed in a proportion of 1 part cement to 3 parts sand and is written as a ratio 1:3. It does not matter what is used to measure the cement and sand as long as they both are measured by the same item, e.g. 1 wheelbarrow of cement and 3 wheelbarrows of sand, or 1 bag of cement and 3 bags of sand. Concrete for paths needs a proportion of 1 part cement, 2 parts sand and 3 parts coarse aggregate: a ratio of 1:2:3 (1 wheelbarrow of cement, 2 wheelbarrows of sand and 3 wheelbarrows of coarse aggregate).

Ratios should be treated the same way as fractions. What you do to one side, you must do to the other. A large ratio can be cancelled down into its simplest term: e.g. 8:4 is the same as 2:1 because both sides are divisible by 4.

■ Calculating a ratio

To calculate a ratio, you must:

1 Calculate the total proportional parts,
    e.g. 2:3 has 5 total parts (2 + 3 = 5).

2 Divide the quantity by the total proportional parts,
e.g. 10 g ÷ 5 = 2 g per part.

3 Multiply the quantity per part by each of the ratios,
e.g. 2:3 is 2 × 2 g : 3 × 2 g = 4 g : 6 g.

EXAMPLE 11

(a) Divide 500 g of sugar in the ratio 7:3.

Total parts	7 + 3 = 10
Quantity per part	500 g ÷ 10 = 50 g
Multiply out	7 × 50 g : 3 × 50 g = 350 g : 150 g

Answer = 350 g:150 g

(b) Divide £275 in the ratio 10:8:7.

Total parts	10 + 8 + 7 = 25
Quantity per part	£275 ÷ 25 = £11
Multiply out	10 × £11 : 8 × £11 : 7 × £11 = £110 : £88 : £77

Answer = £110 : £88 : £77

You can double check your answer by adding the £s together.

£110 + £88 + £77 = £275

■ Comparing ratios

You would use the skill of comparing ratios in situations like telling which of two liquids is the stronger. Concentrations of fluids are almost always expressed as a ratio therefore you must be able to compare ratios.

Comparison of ratios is treated in a similar way to comparison of fractions.

EXAMPLE 12

(a) Compare the ratio 1:20 and the ratio 7:20.

If both right-hand sides are the same number then the more concentrated ratio is the ratio with the largest left-hand side.

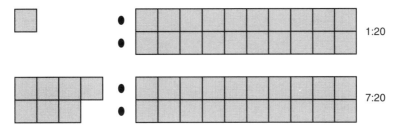

Fig 3.6

Answer = 7:20

(b) Compare the ratio 7:20 and the ratio 7:13.

If both left-hand sides are the same number then the more concentrated ratio is the ratio with the smallest right-hand side.

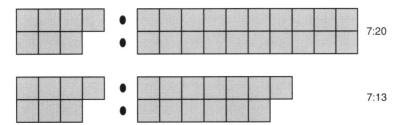

Fig 3.7

Answer = 7:13

(c) Compare the ratio 1:20 and the ratio 7:8.

If the ratios are different on both sides you must find a common right-hand side.

Do not forget, whatever you do to the right-hand side of a ratio you must do the same to the left-hand side.

1:20 becomes 2:40 and 7:8 becomes 35:40
$1 \times 2 : 20 \times 2$ $7 \times 5 : 8 \times 5$

It is now possible to see that 35:40 is a stronger concentration than 2:40.

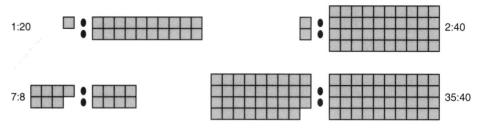

1:20   2:40

7:8   35:40

Fig 3.8

Exercise 3.4

1 Divide the following into the ratios shown:

    (a) £350 in the ratio 2:3   (d) 660 cm in the ratio 19:11

    (b) 1000 km in the ratio 13:12   (e) 282 ml in the ratio 27:20

    (c) 155 g in the ratio 3:2

2 Divide the following into the ratios shown:

    (a) 700 g in the ratio 2:3:2   (d) £5946 in the ratio 2:1:3

    (b) 243 km in the ratio 11:9:7   (e) 1.5 m in the ratio 17:9:4

    (c) 128 litres in the ratio 8:5:3

3 Which of the following ratios is the more concentrated?

    (a) 2:5 and 3:5   (d) 5:7 and 3:11

    (b) 4:5 and 4:9   (e) 2:25 and 3:27

    (c) 1:2 and 2:3

Use fractions, decimal fractions, percentages and ratios to describe situations

Any of the four can be used when describing the results of a survey or questionnaire. Fractions can be used when describing: the proportion of customers spending over £50 in a shop; the proportion of telephone calls made from an office which are long distance and made during peak charge times. Decimal fractions can be used when describing: the dimensions of an object; the amount of fluids in a container. Percentages can be used when describing: the proportion of time a computer network is failing to work; the proportion of people visiting a National Park during the winter period at weekends rather than weekdays. Ratios can be used when describing the average teacher-student ratio in teaching sessions in college departments.

EXAMPLE 13

Forty people were interviewed about a certain washing powder. Twenty-four people use it all the time, 6 people use it occasionally and 10 people never use it.

(a) 24 out of 40 as a fraction:

$$\frac{24}{40} = \frac{3}{5}$$

(b) 6 out of 40 as a decimal:

$$6 \div 40 = 0.15$$

(c) 10 out of 40 as a percentage:

$$\frac{10}{40} \times 100 = 25\%$$

(d) Express the survey information as ratios.

24 people out of 40 becomes 24:16

24 people use the powder all the time and 16 do not; or for every 3 people that use the powder all the time 2 do not.

24:16 cancelled down becomes 3:2

6:34 cancelled down becomes 3:17

10:30 cancelled down becomes 1:3

(e) Express the information as a ratio of use all the time, use occasionally, use never.

24:6:10 cancelled down becomes 12:3:5

Answers =   (a) $\frac{3}{5}$ people interviewed use the washing powder all the time.

(b) 0.15 people interviewed use the washing powder occasionally.

(c) 25% people interviewed never use the washing powder.

(d) 3:2: for every 3 people interviewed that use the washing powder all the time 2 do not.
3:17: for every 3 people interviewed that use the washing powder occasionally 17 do not.
1:3: for every 1 person interviewed that never uses the washing powder 3 do use it.

(e) 12:3:5

Exercise 3.5

1 The price of a train ticket goes up from £28 to £35. What is the percentage increase?

2 A theatre sold 375 tickets for an evening performance. If the theatre seats 600, what fraction is occupied?

3 A company has a budget for redecoration of £10 000, but spends £10 700. By what percentage did it overspend?

4 500 employees were surveyed about the standard of hygiene in the staff canteen. 375 employees thought the standard to be above average. Express this as a decimal.

5 12 employees out of 76 were consistently late for work. 54 employees were never late for work and the remainder were only occasionally late. What percentage of employees are (a) always late; (b) never late?

6 A small chain of supermarket employs 200 males and 500 females. Express this as a ratio of males : females.

7 15 ml of a medicine has to be mixed with 30 ml of water. Express this as a ratio of medicine : water.

Calculate with fractions, decimal fractions, percentages and ratios

The exercises in the last section have given you the opportunity to practise calculation with numbers. Ideally you should calculate with fractions, decimal fractions, percentages and ratios to solve problems.

Exercise 3.6

1 A clothes store gives a 15% discount to students. Helen sees a dress for £25. How much does she have to pay for the dress once the discount is taken into consideration?

2 A personal computer system costs £1585 plus VAT at 17.5%. How much does the computer system cost altogether?

3 A Japanese car manufacturer made 1 500 000 car parts in 1994 and 1 800 000 in 1995. What is the percentage increase in output?

4 Abracadabra nursery school has a teacher : child ratio of 1:6 and Izzie Bizzie nursery school has a teacher : child ratio of 2:11. Which nursery school has less children per teacher?

5 The rent of a flat costs £35 per week. If the cost is increased by 9%, what is the new rental charge?

6 An employee receives a Christmas bonus of £90. It represents $\frac{6}{835}$ of his annual salary. Calculate his annual salary.

7 A map has a scale of 2.5 cm : 5 km. The measurement, on the map, between two towns is 9 cm. What is the distance in km?

8 A store offers a cash discount of 12%. If £365 worth of goods are bought with cash, how much will be saved?

9   Three employees are to share a bonus of £1000. Their annual salaries are in a ratio of 13:11:8 and the bonus is to be paid in the same ratio. How much will each receive?

10 A company employs 500 people. $\frac{7}{10}$ of the workforce are female. How many female employees does the company have?

11 A building firm has a contract to increase the size of the local library by 3.5 times. If the present buildings are 37 500 m^3 what will be the size of the extension?

12 A store sells two brands of peas. Brand A weighs 230 g and costs £1.15 and brand B weighs 250 g and costs £1.25. Which brand is better value for money?

13 A children's home spends $\frac{1}{2}$ of its income on food, $\frac{1}{3}$ on clothes and $\frac{1}{8}$ on electricity. If the income for the year was £24 120, how much was spent on each of the three items?

14 A greengrocer received a delivery of fruit; of which 8% was rotten. If he was able to sell 552 kg, what was the original weight of fruit?

4 Use formulas for calculations

Simple formulas expressed in words

This mathematical technique involves basic algebra. Here you will only be required to use simple algebraic formula.

EXAMPLE 1

If a book on child care is priced at £7 and a book store sells 23 copies, what is the total revenue from the book?

Total revenue is calculated by:

Total revenue = Price × Quantity

Substitute price and quantity into the formula:

$$\text{Total revenue} = \text{Price} \times \text{Quantity}$$
$$= £7 \times 23$$
$$= £161$$

Exercise 4.1

The following exercise uses these formulas:

$$\text{Total cost} = \text{Fixed cost} + (\text{Variable cost} \times \text{Quantity})$$

$$\text{Net profit} = \text{Gross profit} - \text{Expenses}$$

$$\text{Stock available} = \text{Opening stock} + \text{Purchases}$$

$$\text{Gas bill} = \text{Fixed charge} + (\text{Cost per unit} \times \text{Number of units used})$$

$$\text{Average cost per mile} = \frac{\text{Cost of petrol}}{\text{Number of miles travelled}}$$

$$\text{Net wage} = \text{Gross wage} - \text{Deductions}$$

$$\text{Volume of water} = \text{Inflow} - \text{Loss through evaporation}$$

1 What is the total cost of producing 500 items, if the fixed cost is £10 000 and the variable cost is 35p?

2 Gross profit for Head and Sons is £505 600. If expenses totalled £17 850 what is the net profit?

3 The Shoe Shop holds 564 pairs of shoes in stock. If 75 pairs have recently been purchased from the wholesaler, what is the stock available?

4 A hospice used 580 units of gas in a quarter. There is a fixed charge of £13.50 and a cost per unit of 56.8p. Calculate its total gas bill.

5 A salesman travelled 27 500 miles in one year. If the petrol cost for the year was £2350, what is the average cost per mile?

6 A man earns £980 per month. He pays £205 income tax and £55 pension. What is his net wage?

7 10 182 litres of water are passed through a series of pipes on a daily basis. It is known that 1350 litres are lost through evaporation. What is the volume of water left at the end of each day?

Simple formulas expressed in symbols

Many physical, chemical and biological formulas use simple symbols instead of words. To find the voltage of electricity the formula is $V = I \times R$. To calculate speed, the formula $S = \frac{d}{t}$ is used. To find the simple interest earned on money invested the formula is $I = P \times R \times N$.

Some common Greek symbols used in mathematics appear in Table 4.1.

Table 4.1

Symbol	Name	Common uses
α	alpha	Statistics
β	beta	
γ	gamma	
δ	delta	Differential equations
Δ	capital delta	Change
ε	epsilon	
θ	theta	Angles
λ	lambda	Linear variable
μ	mu	
ξ	xi	
π	pi	Constant 3.142
ρ	rho	
σ	sigma	
Σ	capital sigma	Summation
ϕ	phi	
χ	chi	
φ	psi	
ω	omega	Change of an angle
Ω	capital omega	Ohms

EXAMPLE 2

(a) £500 is invested in a bank for 3 years. If the rate of interest is 6% how much interest will the investment earn?

Use the formula $I = P \times R \times N$

where I = Interest (amount of money earned or paid)   = ?
        P = Principal (amount of money invested or borrowed)   = £500
        R = Rate of interest (e.g. 10%) expressed as a decimal (e.g. 0.1)   = 0.06
        N = Number of years of investment or loan   = 3

Substitute the numbers in place of the letters in the formula and multiply.

$I = £500 \times 0.06 \times 3$

$I = £90$

Answer = £90 interest was earned.

(b) A car travels 205 km in 5 hours. What is its average speed?

Use the formula $S = \frac{d}{t}$

where S = Average speed   = ?
       d = Distance (in km)   = 205
       t = Time (in hours)   = 5

Substitute the numbers in place of the letters in the formula and divide.

$S = \frac{205}{5}$

$S = 41$

Answer = average speed is 41 km/h.

Exercise 4.2

The following exercise uses these formulas:

Electricity

$V = I \times R$ where V = Volts
                     I = Current in amperes
                     R = Resistance in ohms

Answer in volts

Simple interest

$I = P \times R \times N$ where I = Interest (amount of money earned or paid)
                     P = Principal (amount of money invested or borrowed)
                     R = Rate of interest (e.g. 10%) expressed as a decimal (e.g. 0.1)
                     N = Number of years of investment or loan

Answer in £s

Average speed

$S = \frac{d}{t}$ where S = Average speed
                d = Distance (in km)
                t = Time (in hours)

Answer in km/h

Density of an object/liquid

$P = \frac{m}{v}$ where P = Density
                m = Mass
                v = Volume

Answer in g/cm³

Temperature

$F = \frac{9}{5}C + 32$ where F = Fahrenheit
                     C = Celsius

Answer in °F

1 A sales representative travels 176 km, from London to Birmingham, in four hours. What is his average speed?

2 A manufacturer wishes to identify the voltage of a battery cell. If the current is 0.6 amperes and the resistance is 5 ohms what is the voltage?

3 A Local Authority invests £7000 at 8% interest for three years. How much interest will it earn?

4 A student borrows £2000 at 15% simple interest for four years. How much interest will s/he pay?

5 A manufactured liquid has a boiling point of 50 °C. What is its boiling point in °F?

6 A dentist purchases a bottle of mercury with a net weight of 272 g and a volume of 20 cm^3. What is its density?

5 Using powers and roots

Powers of numbers

A power of a number increases that number by proportions.

Fig 5.1 shows that the number 2 doubles in size with every increase in the 'power', the number 3 will triple in size and the number 4 will quadruple in size.

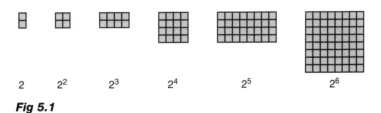

| 2 | 2^2 | 2^3 | 2^4 | 2^5 | 2^6 |

Fig 5.1

Fig 5.2 shows that when a number is multiplied by itself it forms a square. The number is said to be 'squared'. For example 4×4 is 4 squared or 16. 16 is the square of 4.

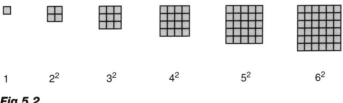

| 1 | 2^2 | 3^2 | 4^2 | 5^2 | 6^2 |

Fig 5.2

When a number is multiplied by itself, and then by itself again, it is said to be 'cubed'. For example $4 \times 4 \times 4$ is 4 cubed or 64. 64 is the cube of 4.

The fourth power of 4 is $4 \times 4 \times 4 \times 4 = 256$.

■ Calculating powers

EXAMPLE 1

(a) $2^2 = 2 \times 2$

$\quad = 4$

(b) $2^5 = 2 \times 2 \times 2 \times 2 \times 2$

$\quad = 32$

(c) $5^2 = 5 \times 5$

$\quad = 25$

(d) $6^3 = 6 \times 6 \times 6$

$\quad = 216$

(e) $9^4 = 9 \times 9 \times 9 \times 9$

$\quad = 6561$

(f) $12^8 = 12 \times 12 \times 12 \times 12 \times 12 \times 12 \times 12 \times 12$

$\quad = 429\ 981\ 696$

It is not always easy to calculate a large power, like the one shown in Example 1(f). It is possible to type the number and multiplication into a calculator over and over again, but a 'power button' on a scientific calculator is the better way.

EXAMPLE 2

A Casio calculator generally shows the power button as X^y and a Texas Instruments calculator generally shows it as Y^x.

Locate the power button on a scientific calculator and calculate 3^{14}.

Casio type calculator	Texas Instruments type calculator
3 X^y 14 =	3 Y^x 14 =

Try to remember the sequence: 3 to the power 14 = OR 3 power button 14 =

Answer = 4 782 969

Roots of numbers

The root of a number is the inverse (opposite) of the power of the number. For example: square root, cube root. Where 4 squared is 16, the square root of 16 is 4, and 4 cubed is 64, the cube root of 64 is 4.

EXAMPLE 3

We have to be careful with the square root of a number as it can be positive (+) or negative (–).

The square root of 16 is + 4 or – 4

To check: $(+ 4) \times (+ 4) = 16$ and $(– 4) \times (– 4) = 16$

The $\sqrt{}$ sign is probably very familiar to you. This sign is used to indicate the positive square root of a number.

EXAMPLE 4

$\sqrt{64}$ = square root of 64 = 8

$\sqrt[3]{216}$ = cube root of 216 = 6

$\sqrt[4]{16}$ = fourth root of 16 = 2

It is possible to work out the root of a number by using a scientific calculator. The root button is usually the shift or inverse of the power button. A Casio calculator generally shows the root button as $X^{1/y}$ and a Texas Instruments calculator generally shows it as $Y^{1/x}$. Locate the power button on a scientific calculator and calculate $\sqrt[5]{16807}$

Casio type calculator	Texas Instruments type calculator
16807 shift $X^{1/y}$ 5 =	16807 INV $Y^{1/x}$ 5 =

Try to remember the sequence: 16807 root 5 = OR 16807 shift, root button 5 =

Answer = 7

Exercise 5.1

1 Mentally calculate the following powers:

(a) 3^3 (b) 2^5 (c) 10^6 (d) 9^2 (e) 3^4

2 Mentally calculate the following roots:

(a) $\sqrt{9}$ (b) $\sqrt{16}$ (c) $\sqrt[3]{125}$ (d) $\sqrt[3]{64}$ (e) $\sqrt{49}$

3 Use a scientific calculator to calculate the following powers:

(a) 3^{12} (b) 5^7 (c) 6^8 (d) 9.11^6 (e) 0.08^9

4 Use a scientific calculator to calculate the following roots:

(a) $\sqrt{0.0144}$ (b) $\sqrt[3]{64\ 000}$ (c) $\sqrt[4]{6561}$

(d) $\sqrt[5]{16\ 807}$ (e) $\sqrt[6]{729}$

Using powers to calculate compound interest and depreciation

The compound interest and depreciation formulas are the ideal way to practise using powers in formulas. The compound interest formula is $A = P(1 + r)^n$

The depreciation formula is $A = P(1 - r)^n$

EXAMPLE 5

£500 is invested at 7% compound interest for 3 years.

The long-hand sequence is:

	£	
Principal sum invested	500.00	
Plus interest received in year 1	35.00	(7% of £500)
Value of investment at the end of year 1	535.00	
Plus interest received in year 2	37.45	(7% of £535)
Value of investment at the end of year 2	572.45	
Plus interest received in year 3	40.07	(7% of £572.45)
Value of investment at the end of year 3	612.52	

The compound interest formula is $A = P(1 + r)^n$

where A = Amount accrued
         P = Principal (a sum of money either invested or borrowed)
         r = Interest rate (expressed as a decimal)
         n = Number of years

Using the power button on a calculator you will be able to arrive at the answer in a much shorter time than by using the long, sequence method.

So if £500 is invested at 7% compound interest for three years we have:

$A = 500 (1 + 0.07)^3$

$A = 500 \times 1.07^3$ (1.07^3 is $1.07 \times 1.07 \times 1.07$)

$A = 500 \times 1.225043$

$A = 612.5215$

Answer = £612.52

To work this out on a scientific calculator you need to find an X^y or Y^x button. Use the following sequence:

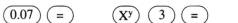
    ___   ___ ___

Exercise 5.2

1 £3500 is invested at 9% compound interest. How much is the investment worth after three years?

2 £1500 is borrowed for four years at 12% compound interest. How much has to be paid back at the end of the period?

3 £15 500 is invested at $5\frac{1}{2}$ % compound interest. How much is the investment worth after 12 years?

4 How much interest is to be paid on a loan of £700 for two years at 11% compound interest?

5 How much interest will I get after seven years if I invest £950 at 8% compound interest?

You may have noticed that the depreciation formula is just a slight variation on the compound interest formula. The depreciation formula is:

$$A = P(1-r)^n \text{ (the + has changed to -)}$$

where A = Value of asset at end of period
       P = Original value of asset
       r = Depreciation rate (expressed as a decimal)
       n = Number of years

Exercise 5.3

1 The value of a machine depreciates at 20% p.a. If it cost £5500 when new, calculate its value after five years.

2 A van cost £15 000 when new. If it depreciates at 12% p.a. what value will it be at the end of five years?

3 A company buys a machine for £30 000 and estimates its life to be ten years. If depreciation is calculated at 14% what will be the value of the machine at the end of the ten years?

4 A machine that has been depreciated by 8% p.a. over five years is now worth £300. What was the original value of the machine?

5 A machine is worth £5000 after five years. If it was depreciated at 16% p.a. what was the original cost of the machine?

6 Probability

Probability scale

Probability is a measurement of how likely it is an event will occur.
Events have different probabilities from 0 to 1.

- A probability of 0 means the event cannot happen (e.g. flying unaided).
- A probability of 1 means the event will happen (e.g. dying).

For events which may or may not happen, the probability lies somewhere between 0 and 1. This is shown in Fig 6.1.

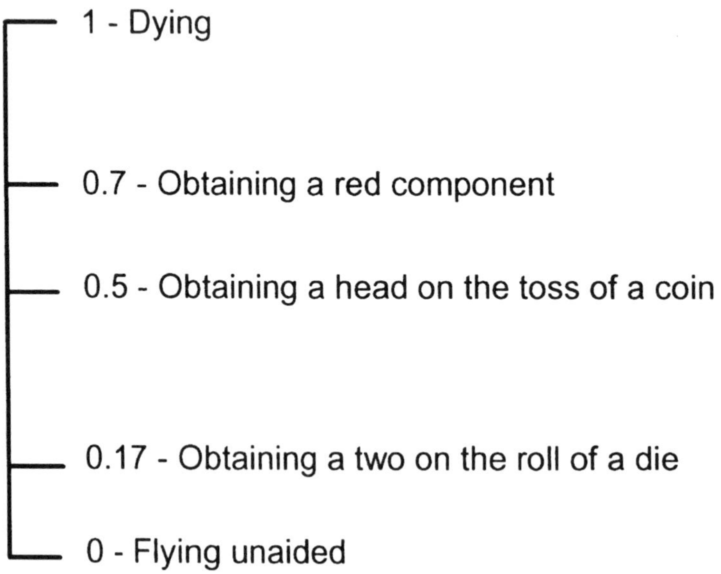

1 - Dying

0.7 - Obtaining a red component

0.5 - Obtaining a head on the toss of a coin

0.17 - Obtaining a two on the roll of a die

0 - Flying unaided

Fig 6.1

Probability formula

The probability of an event occurring is defined as

$$P \text{ (the event)} = \frac{\text{total number of favourable outcomes}}{\text{total number of possible outcomes}}$$

Using probability to describe situations

EXAMPLE 1

(a) What is the probability of a thrown die landing on 4?

$$P \text{ (die landing on 4)} = \frac{1}{6} \quad \frac{\text{total number of favourable outcomes}}{\text{total number of possible outcomes}}$$

(b) A box contains 3 blue components and 7 red components. What is the probability of picking out (i) a blue component and (ii) a red component?

(i) $P \text{ (blue component)} = \dfrac{3}{10}$ $\quad \dfrac{\text{total number of favourable outcomes}}{\text{total number of possible outcomes}}$

$$= 0.3$$

(ii) $P \text{ (red component)} = \dfrac{7}{10}$ $\quad \dfrac{\text{total number of favourable outcomes}}{\text{total number of possible outcomes}}$

$$= 0.7$$

Exercise 6.1

1 What is the probability that a tossed coin will land on tails?

2 What is the probability that a rolled die will show:

(a) a 3   (b) an odd number   (c) a score more than 4?

3 A box contains 12 tennis balls and 13 cricket balls. What is the probability of selecting:

(a) a cricket ball   (b) a tennis ball?

4 A manufacturer of soft drinks regularly checks to see if the machines are filling the bottles to the top. On one check it was found that out of a crate of 24 bottles 4 were underfilled. An inspector selects one bottle at random from the crate. What is the probability that the bottle is underfilled?

5 A mail order company experiences difficulty with customer orders. It was found that out of 500 orders dispatched 15 were the wrong item. What is the probability of a customer receiving the wrong item?

6 A Rolls Royce car is hired out for weddings. The car has been known to break down on 6 occasions out of 100. What is the probability that a bride will be late to her wedding due to the car breaking down?

7 On a production line of 14 000 items, 280 were found to be defective. What is the probability of an item chosen at random being defective?

8 A courier company loses 2 packages out of every 250. What is the probability of an item being lost in transit?

9 A nurse on a ward notes that out of 300 patients 204 remained for 48 hours or less and 96 remained for more than 48 hours. What is the probability that a patient will remain on the ward for more than 48 hours?

10 A large department store offers a delivery service. Out of 125 items delivered 4 were found to be damaged on delivery. What is the probability that a delivered item will be damaged?

■ The addition rule

The total probability of all the possible outcomes of an event must be 1 because one of the events must occur.

EXAMPLE 2

If we take the example of the box which contains 3 blue components and 7 red components, what is the probability of picking out a blue component or a red component?

$$P \text{ (blue component)} = \frac{3}{10}$$

$$P \text{ (red component)} = \frac{7}{10}$$

$$P \text{ (blue component or red component)} = \frac{3}{10} + \frac{7}{10} = 1$$

This extremely obvious example illustrates the addition rule of probability. Another example is the toss of a coin. The probability of tossing a head or a tail on an unbiased coin is 1. The toss of a coin is known to be *mutually exclusive* because you cannot toss a head and a tail at the same time. Likewise you cannot take a red component and a blue component out of the box in one go; the component will be either red or blue. The term *mutually exclusive* means that one event cannot happen at the same time as another event.

■ The multiplication rule

The multiplication rule is used when trying to establish the probability of two or more *independent* events occurring. Events are *independent* when the probability of the first event has no effect on any subsequent events. For example, a first toss of a coin is a head and a second toss of a coin is a head.

EXAMPLE 3

Two boxes of components are arranged as shown in Fig 6.2 on page 60. Box One contains 3 blue components and 7 red components, and Box Two contains 9 green components and 11 yellow components.

What is the probability of obtaining a red component from Box One and a yellow component from Box Two?

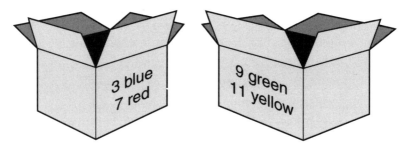

Fig 6.2

P (red component from Box One) $= \frac{7}{10}$

P (yellow component from Box Two) $= \frac{11}{20}$

P (red component and yellow component) $= \frac{7}{10} \times \frac{11}{20} = \frac{77}{200}$

$$= 0.385 \text{ or } 38.5\%$$

Exercise 6.2

1 What is the probability that a tossed coin will land on tails on the first toss and heads on the second toss?

2 What is the probability that a die rolled twice will show:

(a) a 3 and then a 4

(b) an odd number and then an even number

(c) a score more than 4 on both rolls?

3 Box One contains 12 tennis balls and 13 cricket balls, and Box Two contains 2 tennis balls and 3 cricket balls. What is the probability of selecting:

(a) a cricket ball from Box One and a tennis ball from Box Two

(b) a tennis ball from Box One and a cricket ball from Box Two?

4 Out of a crate of 24 soft drinks bottles 4 were found to be underfilled. Out of a second crate of 24 only 2 were found to be underfilled. An inspector selects one bottle at random from each crate. What is the probability that both the bottles are underfilled?

5 A production line produces 14 000 widgets and 15 000 gadgets. On inspection 280 of the widgets and 250 of the gadgets were found to be defective. What is the probability of a widget and a gadget, chosen at random, both being defective?

Probability by experiment

Probability is defined as putting uncertainty into numbers. It is used in industry and business to help solve problems. For example, industry uses probability to determine:

- whether a production line is turning out products at a fast enough rate;
- whether accounts will be paid on time or goods delivered on time;
- whether a product on a production line will be faulty when finished;
- how many letters in a company's internal mailing system get lost.

The probability of an event occurring is based upon experiment. The probability that a coin will land on tails half the number of times it is tossed has been tested by experiment. The probability that a letter will get lost in a company's internal mailing system is also based upon experiment. In this case the experiment is determined by any records kept of people losing letters.

The probability of an event occurring is not always fixed. For example, the probability of a train being a steam train was one in three in 1960. In 1995 it is no longer the case. The setting of insurance premiums has its base in probability experiment and the probabilities regularly change. In the early 1990s insurance premiums rose considerably. This was due to increases in crimes such as car thefts and household burglaries. As the probability of a crime occurring increased, the probability that an insurance company would have to pay out insurance also increased. This therefore resulted in premiums being raised.

Estimating the probability of an event

Recent, valid data is needed to estimate the probability of an event occurring. This is usually collected by observation. When the information is collected, it can be entered into a formula to calculate the probability. The formula is:

$$p(E) = \frac{\text{the number of times the event occurred}}{\text{the number of times the experiment was performed}}$$

EXAMPLE 4

A production manager wishes to know the probability of an item being faulty when made. He checks 850 consecutive items for faults and records 34 faulty items. What is the probability of an item being faulty?

$$p(E) = \frac{\text{the number of times the event occurred}}{\text{the number of times the experiment was performed}}$$

$$p(\text{faulty item}) = \frac{34}{850} = \frac{1}{25}$$

The production manager now knows that 1 in 25 items will be faulty. This will help when filling orders. If an order is placed for 400 items the manager knows that 416 items must be made.

Exercise 6.3

1 An organisation's internal mailing system lost two letters out of 200. What is the probability of a letter going missing?

2 A courier company delivered 54 packages out of 450 late. What is the probability that a delivery will be late?

3 Frisk and Co. had 946 customers out of 1000 pay their account on time. What is the probability that an account will not be paid on time?

4 Tannica and Sons make two products. A production time of 30 minutes is set for Product A and 35 minutes for Product B. The time taken to produce 500 of Product A and 428 of Product B is recorded and analysed. 349 of Product A were produced within the 30 minutes. 296 of Product B were produced within the 35 minutes. Which production run is more efficient, Product A or Product B?

5 A component is made for a speedboat engine. 40 engines are fitted with the component, but two fail to work. What is the probability that an engine fitted with the component will not work?

Part 2
SHAPE, SPACE AND MEASURES

7 Measurement

Metric and imperial measures

In the UK two types of measurement are in use: metric and imperial measures. In 1963 an Act of Parliament specified that the imperial system of weights and measures be replaced with the metric system. This change has taken a long time, and is still not complete. It is only in recent years that petrol filling stations have converted their pumps to litres from gallons. However, road signs still indicate distance in miles. Industry now measures and weighs using the metric system, but there are many people in industry who still use the imperial system.

You should be so familiar with measurement that you can estimate length, weight and capacity in both metric and imperial measures.

Study Tables 7.1 and 7.2 on page 68 and, using a tape measure, set of scales and measuring jug, familiarise yourself with metric and imperial measures.

Once you become familiar with length, weight and capacity you should become familiar with how many smaller units make up the larger units. Tables 7.3 and 7.4 on page 68 will be of help.

Table 7.1 Metric measures

Length	millimetre
	centimetre
	decimetre
	metre
	kilometre
Weight	gram
	kilogram
	tonne
Capacity	millilitre
	centilitre
	litre

Table 7.2 Imperial measures

Length	inch
	foot
	yard
	mile
Weight	ounce
	pound
	stone
Capacity	fluid ounce
	pint
	gallon

Table 7.3 Metric measures

Length	10 millimetres	(mm)	=	1 centimetre	(cm)
	1000 millimetres	(mm)	=	1 metre	(m)
	10 centimetres	(cm)	=	1 decimetre	(dm)
	10 decimetres	(dm)	=	1 metre	(m)
	100 centimetres	(cm)	=	1 metre	(m)
	1000 metres	(m)	=	1 kilometre	(km)
Weight	1000 milligrams	(mg)	=	1 gram	(g)
	1000 grams	(g)	=	1 kilogram	(kg)
	1000 kilograms	(kg)	=	1 tonne	(t)
Capacity	10 millilitres	(ml)	=	1 centilitre	(cl)
	1000 millilitres	(ml)	=	1 litre	(l)
	100 centilitres	(cl)	=	1 litre	(l)

Table 7.4 Imperial measures

Length	12 inches	(in)	=	1 foot	(ft)
	36 inches	(in)	=	1 yard	(yd)
	3 feet	(ft)	=	1 yard	(yd)
	1760 yards	(yd)	=	1 mile	
Weight	16 ounces	(oz)	=	1 pound	(lb)
	14 pounds	(lb)	=	1 stone	(st)
Capacity	20 fluid ounces	(fl oz)	=	1 pint	(pt)
	8 pints	(pt)	=	1 gallon	(gal)

Measuring instruments

It is important when selecting a measuring instrument that you select the most appropriate for the task. It would seem rather silly to use a small 1 litre jug to measure how much water is required to fill a 1000 litre fish tank or a metre ruler to measure the length of a 80 metre corridor. The choice of measuring instrument would also depend upon how accurate a measurement is required. A lab technician is more likely to use a pipette marked with tenths of millilitres rather than a 1 litre jug; a pharmacist will want to use scales marked with grams and a farmer will want to use scales marked with kilograms.

Converting between units

It is important that you can convert from larger units to smaller units and smaller units to larger units. With metric units of capacity you should be able to convert from litres to millilitres; with metric units of mass convert from kilograms to grams; and imperial units of mass from pounds to ounces. You should be able to convert time from hours to minutes to seconds. Other examples include measurement (metres to centimetres) and temperature (°F to °C).

EXAMPLE 1

(a) Convert 40 centimetres into millimetres.

$40 \times 10 = 400$

Answer = 400 mm

(b) How many yards are there in 3 miles?

$3 \times 1760 = 5280$

Answer = 5280 yards

(c) Convert 3400 grams into kilograms.

$3400 \div 1000 = 3.4$

Answer = 3.4 kg

(d) Convert 24 inches into feet.

$24 \div 12 = 2$

Answer = 2 feet

(e) Convert 5 litres into millilitres.

$5 \times 1000 = 5000$

Answer = 5000 ml

(f) Convert 60 fluid ounces into pints.

$60 \div 20 = 3$

Answer = 3 pints

Exercise 7. 1

1 Convert the following to centimetres:

(a) 5 m   (b) 30 mm   (c) 270 mm   (d) 4.27 m

2 Convert the following to millimetres:

(a) 21 cm   (b) 315 cm   (c) 4.6 cm   (d) 5 m

3 Convert the following to feet:

(a) 3 yd   (b) 48 in   (c) $2\frac{2}{3}$ yd   (d) 30 in

4 Convert the following to yards:

(a) 72 in   (b) 12 ft   (c) 2 miles   (d) 108 in

5 Convert the following to metres:

(a) 500 cm   (b) 3 km   (c) 18 000 mm   (d) 4.87 km

6 Convert the following to kilometres:

(a) 17 000 m   (b) 500 000 cm   (c) 6400 m   (d) 6 430 000 cm

7 Convert the following to grams:

(a) 2000 mg   (b) 4 kg   (c) 8300 mg   (d) 17.54 kg

8 Convert the following to kilograms:

(a) 7000 g   (b) 25 000 000 mg   (c) 3 t   (d) 670 g

9 Convert the following to ounces:

(a) 2 lb   (b) 3 st   (c) 4 lb 7 oz   (d) 1 st 8 lb

10 Convert the following to pounds:

(a) 64 oz   (b) 5 st   (c) 48 oz   (d) 2 st 2 lb

11 Convert the following to millilitres:

(a) 2 l   (b) 30 cl   (c) 1.45 l   (d) 74 cl

12 Convert the following to litres:

(a) 5000 ml   (b) 200 cl   (c) 3560 ml   (d) 490 cl

13 Convert the following to pints:

(a) 40 fl oz   (b) 8 gals   (c) 50 fl oz   (d) $4\frac{1}{8}$ gals

14 Convert the following to gallons:

(a) 16 pt   (b) 480 fl oz   (c) 44 pt   (d) 720 fl oz

Measurement of time

Time is also a unit of measurement you are required to be familiar with.

1 second (sec)	=	a very short space of time
60 seconds	=	1 minute (min)
60 minutes	=	1 hour (hr)
24 hours	=	1 day

EXAMPLE 2

(a) Add 1 hour 23 minutes to 3 hours 7 minutes.

$$23 + 7 = 30 \text{ min}$$
$$1 + 3 = 4 \text{ hr}$$

Answer = 4 hr 30 min

(b) Subtract 1 hour 5 minutes from 4 hours 18 minutes.

$$18 - 5 = 13 \text{ min}$$
$$4 - 1 = 3 \text{ hr}$$

Answer = 3 hr 13 min

(c) Add 2 hours 37 minutes to 3 hours 55 minutes.

$$37 + 55 = 92 \text{ min} = 1 \text{ hr } 32 \text{ min}$$
$$2 + 3 + 1 = 6 \text{ hr}$$

Answer   = 6 hr 32 min

Exercise 7.2

Estimate your answer before calculating.

1 Add the following times:

(a) 1 hr 30 min + 2 hr 20 min
(b) 55 sec + 1 min 5 sec
(c) 13 hr 38 min + 5 hr 53 min
(d) 2 days 3 hr + 23 hr

2 Subtract the following times:

(a) 5 hr 50 min – 3 hr 20 min
(b) 3 days – 12 hr
(c) 1 min 15 sec – 45 sec
(d) 2 hr 40 min 30 sec – 1 hr 22 min 13 sec

Compound measures

Some people who drive high performance cars try to attain a speed of 100 miles per hour (mph), whereas the speed limit on a British motorway is 70 mph. A recreation centre will try to attain a fast flow of water when filling the swimming pool. This is measured in litres per minute. A shop may be interested in assessing customers per hour. A typesetter may be concerned about words per square inch.

It is important to remember when measuring rate of change that two or more properties are involved.

EXAMPLE 3

(a) Average speed is calculated by distance over time. This can be km/h, mph, m/s.

    (i)   An object falls 50 metres in 5 seconds. What is its average speed?

            50 metres ÷ 5 seconds = 10 m/s

    (ii)   A train travels 460 miles in 4 hours. What is its average speed?

            460 miles ÷ 4 hours = 115 mph

    (iii)   A car travels 51 km in half an hour. What is its average speed?

            51 km ÷ 0.5 hour = 102 km/h

(b) The density of an object or liquid is calculated by mass over volume. This is measured in g/cm^3.

    A manufacturer has developed a new compound with mass of 500 g and volume of 250 cm^3. Find its density.

    500 g ÷ 250 cm^3 = 2 g/cm^3

(c) Movement of liquids is calculated by capacity over time. This is measured in l/min.

    A pool requires 300 000 litres of water and takes 5 hours to fill. How fast is the water moving? (Remember to convert 5 hours to 300 minutes).

    300 000 ÷ 300 = 1000 l/min

Exercise 7.3

1 A car petrol tank holds 45 litres of petrol. If it takes 4.5 minutes to fill the tank how fast is the pump dispensing the petrol?

2 A tank holds 1 000 000 litres of processed sewage. Its contents are pumped out to sea in 5.5 hours. How fast is the sewage water moving?

3 A conveyor belt in a factory moves the items 12 m in 2 minutes. What is the average speed of the conveyor belt in m/s?

4 A nurse cycles 8 km to work in 45 minutes. What is her average speed in km/h?

5 What is the density of 225 cm^3 of medicine weighing 45 g?

6 A shop is open for a total of 48 hours per week. In one week the shop had 1200 customers. On average how many customers per hour was that?

7 A typesetter manages to get a 3000-word story into 500 in^2. How many words per square inch is this?

8 Perimeter, area and volume

Perimeter

Perimeter is the distance around a shape. You need to know how to calculate the perimeter of a square, a rectangle and a triangle.

EXAMPLE 1

Calculate the perimeter of the badminton court shown in Fig 8.1 using imperial units.

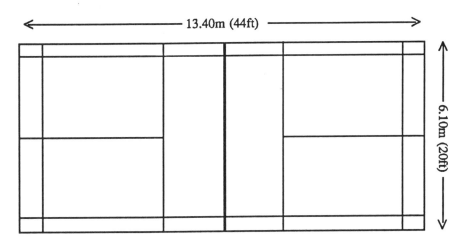

Fig 8.1 A badminton court

The perimeter of a rectangle is:

$(2 \times \text{length}) + (2 \times \text{width})$

or

length + width + length + width.

Therefore the perimeter of the badminton court in Fig 8.1 is:

$(2 \times 44 \text{ ft}) + (2 \times 20 \text{ ft}) = 128 \text{ ft}$

or

44 ft + 20 ft + 44 ft + 20 ft = 128 ft

Exercise 8.1

1 Find the perimeter of each shape in Fig 8.2.

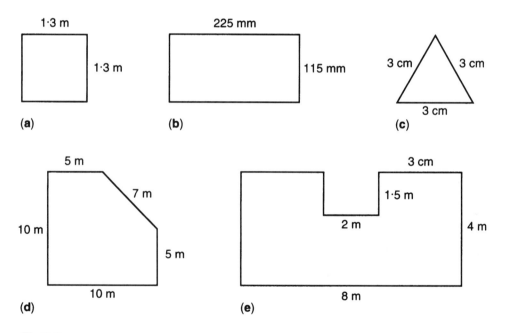

Fig 8.2

2 A lounge room is 5 m × 3 m with a chimney breast of 0.5 m deep by 1.5 m wide. The owner has contracted you to put up a 15 cm wide wallpaper border around the room. What length of border will you require?

3 A farmer wishes to build a wire fence around a field 2206 m × 3673 m. How much fencing will the farmer require?

4 A long distance runner wishes to use the local park for week-day training. If the runner wishes to run 6000 m per day and the park is 940 m × 560 m, how many circuits of the park must be completed to reach this target?

5 A triangular bandage needs a border to stop the bandage from fraying. How much border is needed if the bandage measures 1200 mm × 900 mm × 900 mm?

Area of plane shapes

The area of a plane shape is measured by the number of square units it contains.

Area can be calculated in two ways: using a square grid and using formulas. The use of a square grid will only give an approximate answer whereas using a formula will give a precise answer.

■ Using a square grid

EXAMPLE 2

To calculate the area count the number of squares within the shape.

For the shapes in Fig 8.3 on page 78:

The area of (a) = 25 squares

The area of (b) = 24 squares.

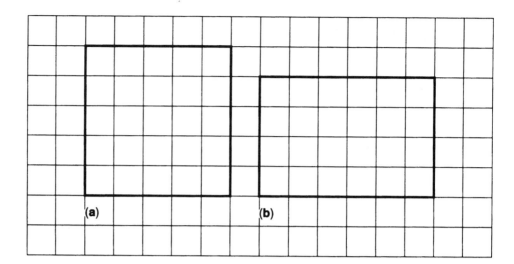

Fig 8.3

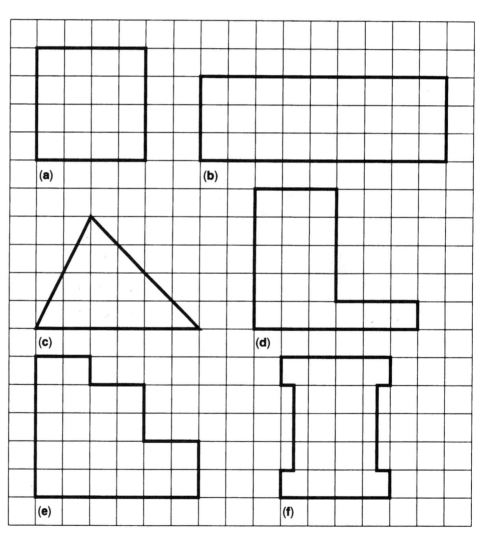

**Fig 8.4
(Scale:
1 square =
1 cm²)**

Exercise 8.2

Calculate an approximate area for the shapes in Fig 8.4.

■ **Using formulas**

Look at the shapes in Fig 8.5 and learn the following formulas for calculating their areas:

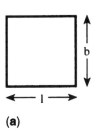
(a)

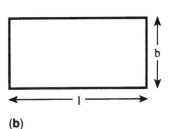
(b)

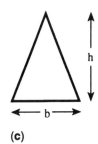
(c)

Fig 8.5

(a) The area of a square $\quad= l \times b$

(b) The area of a rectangle $= l \times b$

(c) The area of a triangle $\quad= \frac{1}{2} b \times h$

EXAMPLE 3

(a) A carpet is 3 m square. What area is it?

Formula for square is $l \times b$.

Answer is $3 \times 3 = 9 \text{ m}^2$.

(b) A sheet of metal is 20 cm by 15 cm. What area is it?

Formula for rectangle is $l \times b$.

Answer is $20 \times 15 = 300 \text{ cm}^2$.

(c) A triangular sign has a base of 60 cm and a height of 70 cm. What area is it?

Formula for a triangle is $\frac{1}{2} b \times h$.

Answer is $\frac{1}{2} \times 60 \times 70 = 2100$ cm^2.

Exercise 8.3

1 A garden is 5 m × 7 m. What area is it?

2 What is the area of a sheet of metal measuring 58 cm × 32 cm?

3 Find the area of a field measuring 507 m × 210 m.

4 Find the area of a triangular bandage that has a base of 1200 mm and a perpendicular height of 600 mm.

5 A triangular shaped clock has a base of 60 cm and a perpendicular height of 70 cm. How much area will it take up on an office wall?

6 The stage of a local theatre measures 45 m by 30 m. A travelling theatre group needs 1200 m^2 for its performance. Is the stage large enough?

7 The blood bank wishes to use a community centre for a blood donor session. Trellis Hall is 5 m by 9 m and Cumberland Hall is 7 m by 8 m. The blood bank has five beds, each bed requiring 6 m^2 of floor area. Which hall would be most suitable?

8 The hold of a cross channel ferry measures 25.2 m by 153 m and cars are parked end to end. If the average car takes up an area of 16.20 m^2, how many cars will the ferry carry?

9 A confectionery company is selling its chocolates in a triangular based box. The base measurements are given as 7.5 cm and 6.5 cm. If a shop allocated 195 cm^2 of shelf space, how many boxes can be displayed at a time?

10 A company makes give way signs. Each sign has a base of 0.6 m and a height of 0.6 m. How many signs can be made out of 2 m^2 of metal?

Volumes of simple solids

The volume of a simple solid is measured by the number of cubic units it contains. You are required to know the formulas for a cube (or cuboid) and a prism. (See Fig 8.6.)

(a) The volume of a cube $= l \times b \times h$

(b) The volume of a cuboid $= l \times b \times h$

(c) The volume of a prism $= \frac{1}{2} b \times h \times l$ or area of cross-section $\times l$

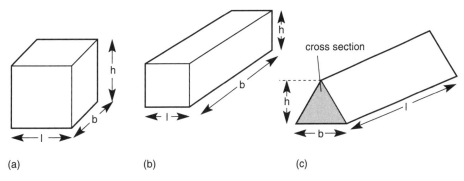

(a)   (b)   (c)

Fig 8.6

EXAMPLE 4

(a) A box is a 3 m cube. What volume is it?

The length, width and height of a cube are the same measurement. The formula for the volume of a cube is:

$l \times b \times h$

Answer is 3 m $\times$ 3 m $\times$ 3 m $= 27$ m^3.

(b) A metal container is 3 m by 6 m by 11 m. What volume is it?

The length, width and height of a cuboid are different measurements. (Two measurements may be the same.)

The formula for the volume of a cuboid is:

$l \times b \times h$

Answer is 3 m $\times$ 6 m $\times$ 11 m $= 198$ m^3.

(c) A prism has a base of 4 cm, a height of 4 cm and a length of 15 cm. What volume is it?

The formula for the volume of a prism is:

half the base × the perpendicular height × length

or

the area of cross-section × length

Answer is 2 cm × 4 cm × 15 cm = 120 cm³.

Exercise 8.4

1 A die is 20 mm × 20 mm × 20 mm. What is its volume?

2 What is the volume of a container lorry that measures 3.5 m × 4.3 m × 11.2 m?

3 Find the volume of a filing cabinet that measures 47 cm × 62 cm × 132 cm.

4 Find the volume of a prism shaped piece of metal. The length measures 3000 mm and the area of the cross-section measures 2550 mm².

5 Find the volume of a triangular shaped box of chocolates that has a base of 30 cm, a height of 17 cm and a length of 5 cm.

6 An office replaces five filing cabinets measuring 50 cm × 60 cm × 140 cm with three roll front cabinets measuring 110 cm × 50 cm × 200 cm. How much more space will the new cabinets require?

7 A storage company owns 320 containers and needs to buy a new warehouse. Each storage container measures 4.6 m × 3.2 m × 8.7 m. What is the minimum size warehouse the company should buy? Give your answer in m³.

8 A jeweller has 2430 mm³ of gold that he wishes to make into charms for a bracelet. If each charm has a length of 4 mm and a cross-sectional area of 4.5 mm², how many charms will he make?

9 A box 30 cm × 32 cm × 57 cm contains 14 books of equal size. What is the volume of each book?

10 From a cuboid piece of stone 60 cm × 60 cm × 90 cm a mason carves a prism shaped table support. The support has a base of 60 cm, a height of 60 cm and a length of 90 cm. What volume of stone has been wasted?

Calculating area and volume with mixed units

The rule to remember when calculating with mixed units is always to convert to the units required in the answer before calculating area and volume.

EXAMPLE 5

(a) A carpet measures 95 cm × 3 m. What is its area in m²?

95 cm = 0.95 m

Answer is 0.95 m × 3 m = 2.85 m².

A common mistake is to convert to cm (answer = 28 500 cm²) and then divide by 100. There are 100 cm in a metre, but 10 000 cm² in 1 m².

1 m × 1 m = 1 m²

This is the same as:

100 cm × 100 cm = 10 000 cm²

Therefore if 1 m² = 10 000 cm²; 28 500 cm² = 2.85 m²

(b) A wooden box measures 50 cm × 70 cm × 2 m. What is the volume in cubic metres?

50 cm = 0.5 m

70 cm = 0.7 m

Answer is 0.5 m × 0.7 m × 2 m = 0.7 m³.

The mistake mentioned for area is also made with volume. Many people divide a cm³ answer by 100 instead of 1 000 000.

50 × 70 × 200 = 700 000 cm³

To express the answer in m³, remember:

1 m × 1 m × 1 m = 1 m³

This is the same as

100 cm × 100 cm × 100 cm = 1 000 000 cm³

Therefore if 1 m³ = 1 000 000 cm³; 700 000 cm³ = 0.7 m³.

Circles

It is not sufficient just to know the formulas when calculating the area and circumference of circles – you must understand and know what each part of the formulas means.

The two formulas are:

Circumference	Pi times the diameter	πd
Area	Pi times the square radius	πr^2

The *circumference* is the distance around the edge of a circle. Pi is the ratio of the circumference of a circle to its diameter and equals 3.142 to three decimal points (3 dp) or $\frac{22}{7}$. This means that the circumference is just over three times longer than the diameter. The *diameter* is the distance across the centre of a circle from one edge to the other. The *radius* is the distance from the centre of a circle to the outer edge. Therefore, the radius is half the length of the diameter (see Fig 8.7).

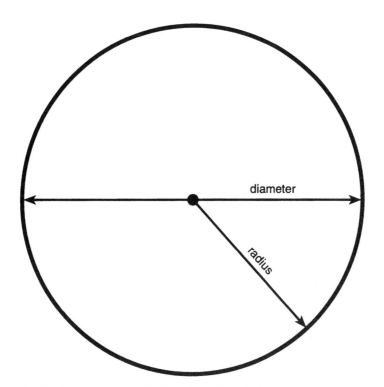

Fig 8.7 The radius and diameter of a circle

■ Circumference of a circle

EXAMPLE 6

(a) A circle has a diameter of 6 cm. What is the circumference?

Formula is πd

Answer is 3.142×6 cm = 18.852 cm.

If you use the pi (π) button on a scientific calculator, the answer will be 18.849556 cm or 18.85 cm (2 dp).

(b) What is the circumference of a circle that has a radius of 2.5 cm?

Formula is πd. Diameter is twice the length of the radius.

Answer is $3.142 \times 2 \times 2.5$ cm = 15.71 cm.

The pi (π) button on a calculator gives an answer of 15.707963 cm or 15.71 cm (2 dp).

(c) A circle has a circumference of 22 m. What is the diameter?

$$3.142 \times d = 22 \text{ m}$$
$$d = \frac{22}{3.142} \text{ m}$$
$$d = 7 \text{ m}$$

It is not sufficient to think you can rely on a calculator to remember pi for you. You should make an effort to remember that pi = 3.142.

■ Area of a circle

EXAMPLE 7

(a) A circle has a radius of 3 m. What is its area?

Formula is πr^2

$3.142 \times 3^2 = 28.26$

or

$3.142 \times 3 \times 3 = 28.26$

Area = 28.26 m^2.

(b) What is the area of a circle that has a diameter of 8 cm?

Formula is πr^2. Diameter divided by 2 = radius. $8 \div 2 = 4$

The radius is therefore 4 cm

$3.142 \times 4^2 = 50.272$

Area = 50.272 cm^2.

(c) What is the radius of a circle if the area is 78.55 cm^2?

$$3.142 \times r^2 = 78.55$$
$$r^2 = \frac{78.55}{3.142}$$

$$r^2 = 25$$
$$r = \sqrt{25}$$
$$r = 5$$

Radius = 5 cm.

Exercise 8.5

1 If the diameter of a circle is 30 mm, what is its circumference?

2 What is the area of a circle that has a radius of 5 cm?

3 Find the circumference and area of a circle that has a radius of 65 mm.

4 The area of a circle is 50.272 m². What is the radius?

5 Find the diameter of a circle if the circumference is 9.426 cm.

6 The diameter of a car wheel is 57 cm. What is the area?

7 A young mother wishes to erect a fence around a garden pond to protect her children. If the pond is 2.5 m in diameter, how much fence will she need?

8 A Town Hall clock has a long hand that measures 75 cm. What are the circumference and area of the clock?

9 A builder knows the circumference of a piece of pipe to be 47 130 mm. What is the diameter?

10 A helicopter landing circle needs to be painted on an airfield. If the circle needs to be a minimum of 153 m², what should the radius be correct to the nearest metre?

Area and volume of combined shapes

EXAMPLE 8

(a) Fig 8.8 shows the dimensions of an athletics track. Calculate the total area.

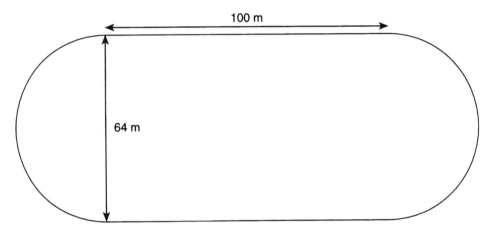

100 m

64 m

Fig 8.8 Dimensions of an athletics track

The track consists of two two-dimensional (2D) shapes: a rectangle and a circle (two semicircles at either end).

The area of the rectangle is

64 m × 100 m = 6400 m^2

The area of the circle is

Radius = 32 m

$3.142 \times 32^2 = 3217.4$ m^2

Area of rectangle + area of circle = Total area

6400 m^2 + 3217.4 m^2 = 9617.4 m^2

Answer is 9617.4 m^2.

(b) Fig 8.9 shows the dimensions of a swimming pool at a local recreation centre. Calculate the total volume.

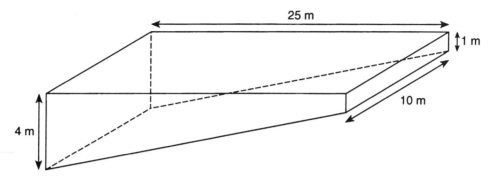

Fig 8.9 Dimensions of a swimming pool

The pool consists of two three-dimensional (3D) shapes: a cuboid and a prism.

The volume of the cuboid is

1 m × 25 m × 10 m = 250 m³

The volume of the prism is

$\frac{1}{2}$ × 3 m × 25 m × 10 m = 375 m³

The volume of the cuboid + the volume of the prism = Total volume.

250 m³ + 375 m³ = 625 m³

Answer is 625 m³.

Volume of a cylinder

The volume of a cylinder is the area of the cross-section × length.

$\pi r^2 h$

EXAMPLE 9

A cylinder has a radius of 20 cm and a height of 65 cm. What is the volume of the cylinder?

The formula is $\pi r^2 h$.

3.142 × 20² × 65 = 81 692

Answer is 81 692 cm³.

Exercise 8.6

1 A baked bean can has a radius of 33 mm and a height of 130 mm. What is the volume of the can?

2 A children's concrete playground is 10 m × 25 m. It is decided to replace the concrete with grass, but leave a 1.5 m wide path all the way round the grass. What will be the area of the pavement?

3 A pipe has an outside diameter of 51 mm and a bore of 48 mm. What is its cross-sectional area?

4 What is the volume of a hot-tub that measures 180 cm in diameter and 92 cm in height?

5 A thin sheet of metal is used to construct an open container 65 cm × 24 cm × 1.2 m. Assuming that the edges do not overlap, calculate the area of the sheet metal used in its construction. Give your answer in both square centimetres and square metres.

6 A piece of wood 1 m × 1.5 m is used to make an occasional table. The table top is circular with a diameter of 55 cm. The support is two rectangles 50 cm × 65 cm slotted together. Calculate the area of wood wasted.

7 A football pitch is 110 m × 75 m. The groundsman uses a roller 1.2 m in diameter and 1.7 m wide to level the pitch. Calculate how many complete revolutions the roller has to make to level the ground.

8 A nurse has two syringes; she must choose the one with the largest volume. Syringe A is 1.5 cm in diameter and 8 cm long, Syringe B is 1.25 cm in diameter and 9 cm long. Which one should she choose?

9 Two-dimensional and three-dimensional diagrams

Everyday shapes

The shapes shown below are common, everyday shapes. It is important you are able to instantly recognise them.

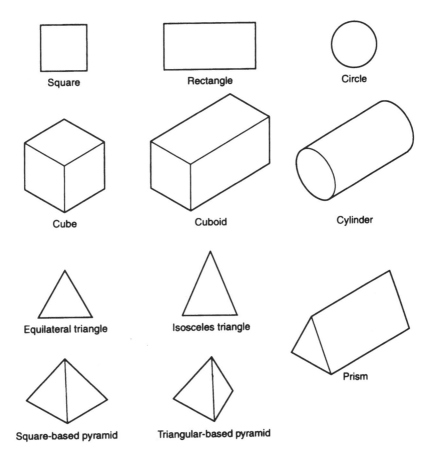

Fig 9.1

From these shapes you should be able to identify everyday objects (or objects common to your vocation). For example:

Square	Chess board, table top, computer floppy disk case
Rectangle	Piece of A4 paper, seat on a park bench
Circle	Clock, dart board, coin, satellite dish, compact disc
Cube	Dice, OXO cube, Rubix cube
Cuboid	Filing cabinet, TV, video
Cylinder	Baked bean can, pen
Equilateral triangle	Give way road sign
Isosceles triangle	Sandwiches you buy from the supermarket in a plastic case
Prism	Roof of a house, Toblerone bar
Square-based pyramid	Church spire, Egyptian pyramids
Triangular-based pyramid	Chocolate sweet

Two-dimensional diagrams of three-dimensional objects

This requires you to:

- draw and recognise nets of common 3D objects
- draw 3D objects
- interpret simple plans and elevations.

■ Recognising and drawing nets

A net is the two-dimensional representation of a three-dimensional object when opened out, e.g. a cereal box.

Exercise 9.1

1 Name the 3D shapes for the nets in Fig 9.2.

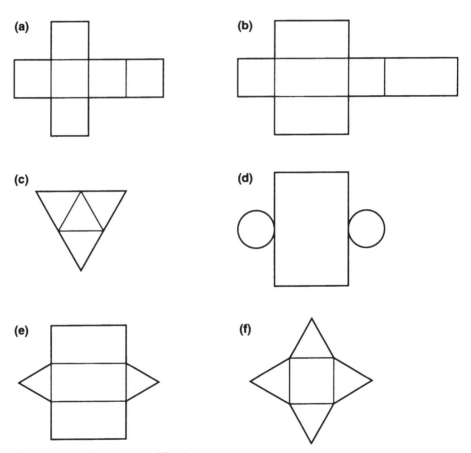

Fig 9.2 Nets for various 3D shapes

2 Draw the nets for the following objects:

(a) a 5 cm square box

(b) a cassette case 7 cm × 11 cm × 1.5 cm

(c) a Toblerone bar

(d) a baked bean can

(e) a triangular-based pyramid

■ Drawing 3D objects

Begin by drawing some of the shapes from the beginning of the chapter. It can be difficult to draw three-dimensional shapes on paper, so isometric grid paper has been provided (see Fig 9.3). Place a plain piece of paper over the isometric grid and use the lines as a guide.

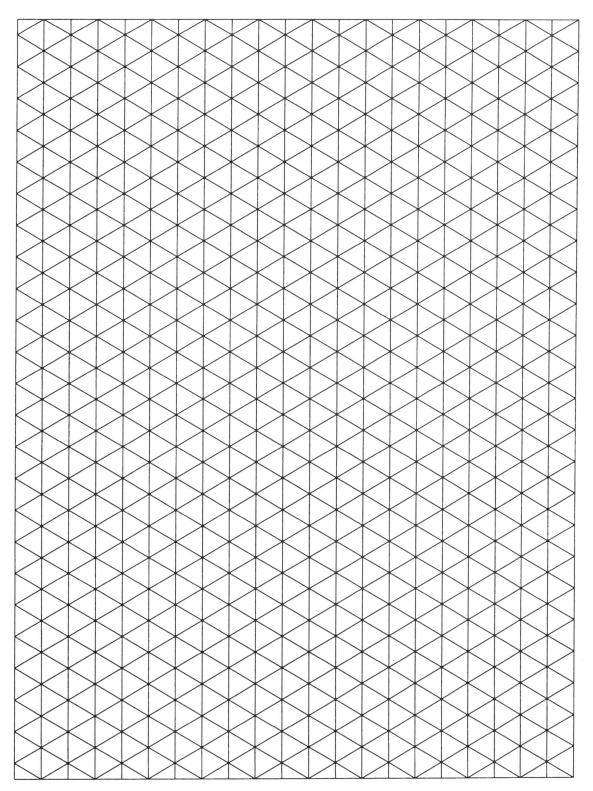

Fig 9.3 Isometric grid

Exercise 9.2

1 Identify the objects in Fig 9.4.

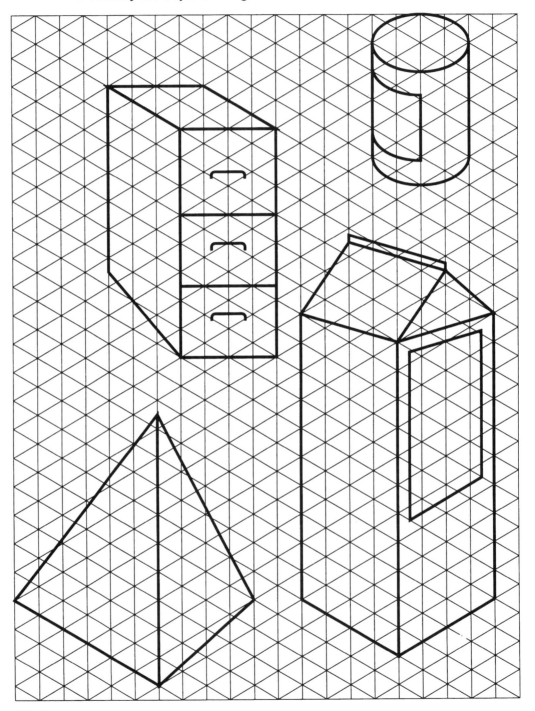

Fig 9.4

2 Draw the following objects:

(a) a filing cabinet with three drawers

(b) an oil drum

(c) a house

Interpret simple plans and elevations

Fig 9.5 is a plan of a kitchen. Learn the symbols for windows and doors (especially the way they open).

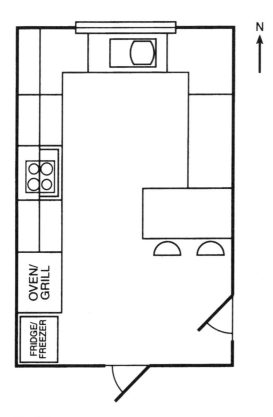

OVEN/GRILL

FRIDGE/FREEZER

N

Fig 9.5

The plan shows the following:

- the hob is halfway down the room on the west wall;
- the window is on the north wall;

- there is a door on the south wall and at the bottom of the east wall;
- there is a breakfast bar on the east wall next to the door;
- the sink is under the window.

Fig 9.6 shows an elevation of part of the west wall in the kitchen.

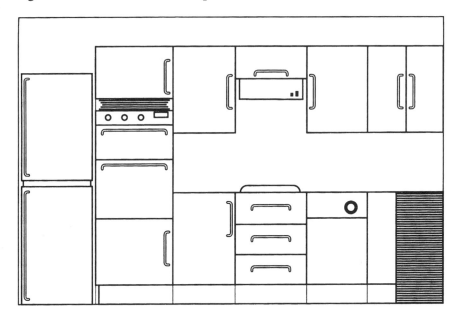

Fig 9.6

The elevation shows the following:

- there is a cooker hood above the hob;
- there is a split level oven in the kitchen;
- the wall cupboards do not reach to the ceiling.

Exercise 9.3

Study the house plans in Fig 9.7 and answer the following questions:

1 What is the largest room on the ground floor?

2 How many windows are there in the master bedroom?

3 What room is over the study?

4 Where does the internal door to the garage go from?

5 What room has a built-in cupboard?

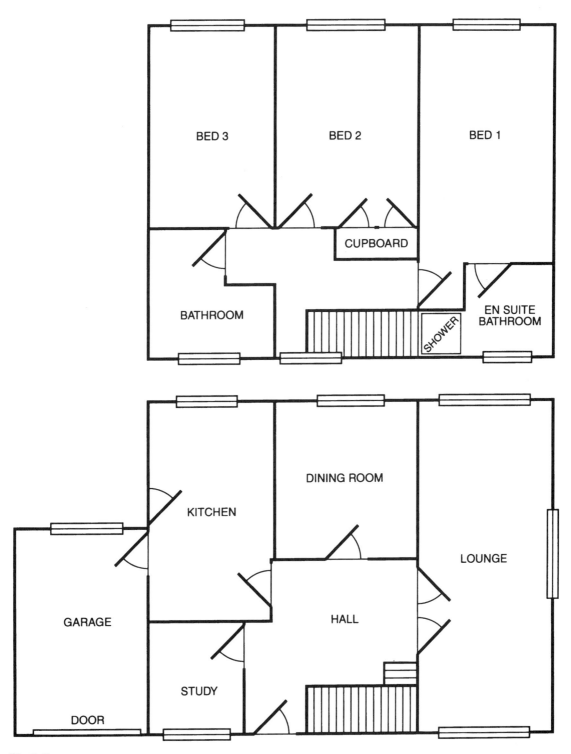

Fig 9.7

10 Network diagrams and network analysis

A network is an interconnected group or system, such as a networked computer system, a road or rail network, a telephone network and the national grid. Network diagrams and network analysis are used to help clarify or solve problems. Networks are made up of a series of nodes (points) and arcs (lines). Managers use sophisticated network analysis to aid in the planning and control of projects.

Network diagrams

EXAMPLE 1

The map shown in Fig 10.1 is of some major towns in the state of Ontario, Canada. A train service connects these towns in the following way:

- Toronto to Ottawa
- Toronto to Sault Ste. Marie
- Ottawa to Sault Ste. Marie
- Ottawa to Fort Albany
- Fort Albany to Thunder Bay
- Fort Albany to Fort Severn
- Fort Severn to St Joseph
- St Joseph to Thunder Bay
- Thunder Bay to Sault Ste. Marie.

Fig 10.1

The network diagram of this information is shown in Fig 10.2.

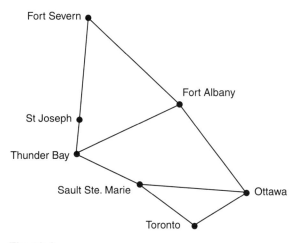

Fig 10.2

A family on holiday wish to travel from Toronto to Fort Severn while changing trains as little as possible. What route should they take?

Answer = Toronto → Ottawa → Fort Albany → Fort Severn (two train changes).

Exercise 10.1

1 Study the town centre road network in Fig 10.3. Some of the roads allow two-way traffic and some allow only one-way traffic. What route must be taken to get from the shopping centre to the market by car?

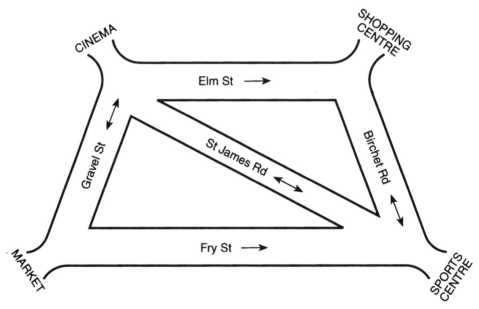

Fig 10.3 Town centre road network

2 Sunny-day Tours offers chartered holidays to Italy. A boat is chartered by the tour operator to take the holiday makers to the towns and cities shown on the map in Fig 10.4. They wish to travel from Naples to Cagliari, then on to Catania and finally on to Lecce. The tourists then wish to travel over land from Lecce to Naples and from Naples to Rome. Construct a network diagram of this information.

3 Mike is a sales representative for an international pharmaceutical company. He works from an office in London. During the first week of every month Mike has to visit a customer in each of the towns shown on the map in Fig 10.5. He wishes to make these visits in as short a distance as possible.

   (a) Using the mileage table shown in Fig 10.6 (on page 104) construct a network diagram of his journey.
   (b) What is his total mileage?

Fig 10.4

Fig 10.5

	Birmingham	Cardiff	Hull	Liverpool	Manchester	Nottingham	Portsmouth
Cardiff	103						
Hull	123	224					
Liverpool	93	169	130				
Manchester	80	172	95	35			
Nottingham	50	153	90	98	63		
Portsmouth	141	142	233	234	221	175	
London	105	167	168	202	185	122	70

Fig 10.6

4 Green Ford has a large dog population and a problem with strays. The town employs a dog catcher, Eric, to visit the local parks each day and capture any stray dogs. The five parks (V, W, X, Y and Z) and the dog compound are shown in Fig 10.7.

(a) Construct a network diagram of all the possible routes Eric can take from the dog compound to each of the parks.

(b) In order to keep costs down Eric must visit the parks in the shortest possible distance. What is this distance?

(c) The B3021 road is closed due to a burst pipe. How much further does he have to travel?

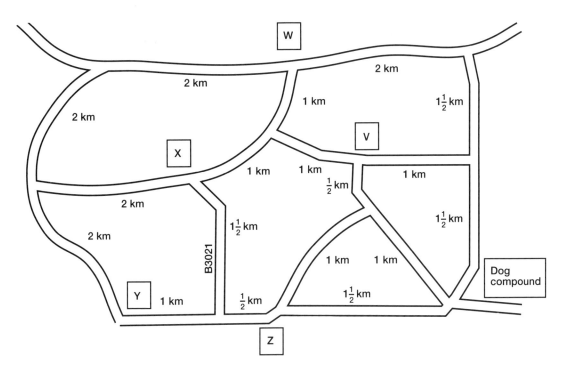

Fig 10.7

Simple network analysis

Simple network analysis gives a pictorial account of activities. It shows which activities need to be completed before others can be begun (*consecutively*). It also shows which activities can be done at the same time (*simultaneously*).

When you are designing a network for analysis note the following:

- An arrow going from left to right represents an activity. This is known as the *arc*.

- The beginning or end of an activity (called the event) is represented by a circle. This is called the *node*.

- A complete network should have a start and a finish.

EXAMPLE 2

A student has six activities to complete in a week.

In Step 1 the activities are identified with a letter and displayed in a table (see Table 10.1). (The activities are not in order of priority.)

Table 10.1

Task	Activity
A	Get statistical information from library.
B	Attend lecture on how to analyse statistical information.
C	Analyse statistical information for Assignment 1, Task 1.
D	Copy up friend's notes as missed last week's lecture.
E	Write a report for Assignment 1, Task 3 using last week's lecture notes.
F	Hand in Assignment 1 for marking.

In Step 2, a third column, Preceding activity, is added to the table. The preceding activity is the activity that comes immediately before the task, e.g. Task B immediately precedes Task C, but not Task F (see Table 10.2).

Table 10.2

Task	Activity	Preceding activity
A	Get statistical information from library.	–
B	Attend lecture on how to analyse statistical information.	A
C	Analyse statistical information for Assignment 1, Task 1.	B
D	Copy up friend's notes as missed last week's lecture.	–
E	Write a report for Assignment 1, Task 3 using last week's lecture notes.	D
F	Hand in Assignment 1 for marking.	C, E

A dash indicates that there is no preceding activity to this task. This task therefore begins at the start of the network. Fig 10.8 shows the network diagram.

(a) What two activities do not rely on any preceding activity?

Answer = The copying up of notes and getting the statistical information from the library.

(b) What is the final activity?

Answer = The handing in of the assignment.

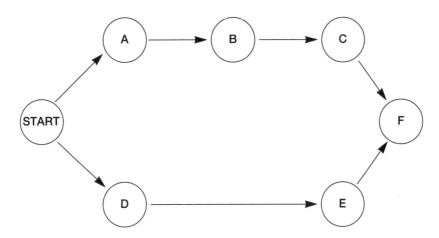

Fig 10.8 Network diagram for Example 2

Exercise 10.2

1 Draw the network analysis diagram for the activity table shown in Table 10.3.

Table 10.3

Activity	Preceding activity
A	–
B	A
C	A, B
D	A
E	B
F	C, D, E
G	F

2 A company has to get a report out as soon as possible. Two secretaries are assigned to assist the manager is achieving this task. Table 10.4 identifies the main activities for this move.

Table 10.4

Task	Activity
A	Take dictation for the report from manager.
B	Type up dictated notes.
C	Photocopy dictated notes.
D	Collect plans for inclusion in the report.
E	Photocopy the plans.
F	Collate report and plans.
G	Check report and plans with manager.
H	Put the report and plans in the post.

Draw a network to represent the activities.

3 A manufacturing company is about to produce a new product. The activities involved are shown in Table 10.5.

Table 10.5

Task	Activity
A	Collect product specification from R&D.
B	Order raw materials.
C	Order new machinery.
D	A few staff go on a training course to use the new machinery.
E	Raw materials arrive at warehouse.
F	Raw materials are checked and stored.
G	New machinery arrives at production plant.
H	New machinery is installed.
I	New machinery is tested.
J	The first manufacturing run is started.
K	New product is tested.
L	Finished product is stored in warehouse.

Produce a prior activity table and draw the network of the activities.

Part 3
HANDLING DATA

11 Collecting data

Data collection plays a very important role in the day-to-day running of an organisation. Some organisations spend hundreds of thousands of pounds in employing a market research company to assess their position in the market-place. Others spend money on management consultants. A management consultant is someone who observes an organisation's employees and suggests ways they can work more efficiently.

There are two main data collection procedures. They are collecting primary data and collecting secondary data.

■ Primary data

Primary data is usually obtained by observing activity or by conducting a survey. A survey can be something very simple such as a show of hands or a more complicated written questionnaire.

■ Secondary data

Secondary data is data which has been collected by another person or organisation. Secondary data can sometimes be very easily accessible, such as statistical data from the Central Statistical Office (a government organisation). Data collected by organisations or companies may not be easily accessible. Company reports and accounts are made public but information such as results of tests on products or research and development proposals are kept very secret.

Survey methods

There are five common survey methods:

1 Show of hands
2 Face-to-face interview
3 Telephone interview
4 Questionnaire
5 Postal questionnaire

The most appropriate method for your survey will depend upon the complexity of the issue you choose.

■ Show of hands

Show of hands is a very easy-to-use method with instant results. The surveyor asks a question and each respondent raises his or her hand. The raised hands are then counted.

Disadvantages

1 Its scope is limited.
2 It is sometimes viewed as being childish.
3 Many people are shy about personal or confidential information.

EXAMPLE 1

A survey of employment patterns was conducted among 20 students. The question was asked, 'If you have a regular part-time job raise your hand.' The raised hands were counted. 15 people out of 20 have a regular part-time job.

■ Face-to-face interview

Many market research companies use face-to-face interviews to survey the public. The interviewers are trained and fairly well paid, which makes this survey method costly.

Advantages

1   The interviewer can show examples of the product, advertisement, article being surveyed.
2   The interviewer can ask more detailed questions or explain difficult questions.
3   The interviewer can watch for exaggerated or obvious lies to answers (e.g. a man wearing an overcoat in the office and stating that the temperature is comfortable).
4   More time can be spent as people do not wish to be impolite and cut short the interview.
5   It is possible to obtain more interviews as many people cannot say no when confronted face to face.

Disadvantages

1   Interviewers can introduce bias into the survey, by the way they ask the questions or the way they record the answers.
2   The method is also time consuming as each interview can take up to 10 minutes.

EXAMPLE 2

A survey was conducted into the taste preference between two similar fizzy drinks. People were asked the following questions:

- 'Which drink do you prefer, A or B?'
- 'Do you regularly buy the drink you have chosen?'
- 'Would you buy it if you were given a money-off voucher?'

The survey resulted in the following:

- 55% of the people surveyed preferred Drink A, 40% preferred Drink B, and 5% did not like either drink.
- 30% of the people surveyed said they bought their preferred drink regularly.
- 80% of the people surveyed accepted a money-off voucher.

■ Telephone interview

The telephone interview is less costly than the face-to-face interview. However, it is easier for someone to decline an interview by replacing the receiver. Other disadvantages are that the interview is limited by time (because of the cost of the call) and limited to people with telephones.

EXAMPLE 3

A survey was conducted into the swimming habits of households. The surveyor began by choosing 20 telephone numbers at random. Each telephone call was started with the surveyor saying, 'I am conducting a survey into the swimming habits of households.' People were then asked the following questions:

- 'Does any member of your family regularly participate in swimming?'
- 'How often do the members of your family participate in swimming?'
- 'Does any member of your family receive swimming lessons?'
- 'Does your family prefer to attend a regular swimming pool or one with a wave machine and water chute?'

The survey resulted in the following:

- 35% of the people surveyed have a family member who regularly participates in swimming.
- The most common answer to how often they participated in swimming was once a week.
- 20% of the families surveyed have a family member who receives swimming lessons.
- 60% of the families surveyed preferred to attend a pool with a wave machine and water chute.

■ Questionnaire

Most questionnaires are given out in stores or magazines.

Advantages

1 A person can answer the questionnaire in his or her own time.
2 Personal questions can be answered in confidence (no one need know who completed the questionnaire).

Disadvantages

1   There is no one to explain difficult questions.
2   The response rate is very poor.
3   The questionnaire has to be well designed to make analysis possible.
4   The questionnaire may be completed by a group of people thus making the information invalid.

■ Postal questionnaire

A postal questionnaire has the same advantages and disadvantages as a questionnaire. The response rate, however, is even poorer (about 2–10%).

EXAMPLE 4

Car owners received the questionnaire shown in Fig 11.1 through the post three weeks after having their car serviced.

| Do you still own the car that was recently serviced by us? | Yes ☐ |
| | No ☐ |

| Are we the dealership that you purchased the car from originally? | Yes ☐ |
| | No ☐ |

Please indicate by ticking the appropriate box how much you agree or disagree with the following statements.

	Agree strongly	Agree slightly	Neither agree nor disagree	Disagree slightly	Disagree strongly
I was attended to within a reasonable length of time	☐	☐	☐	☐	☐
The staff were courteous and helpful at all times	☐	☐	☐	☐	☐
The mechanic understood my requirements	☐	☐	☐	☐	☐
My car was left in safe and competent hands	☐	☐	☐	☐	☐
All my instructions were carried out satisfactorily	☐	☐	☐	☐	☐
The work was carried out to a high standard	☐	☐	☐	☐	☐
It is likely that I shall return for my next service	☐	☐	☐	☐	☐

Fig 11.1 Example of a postal questionnaire

Questionnaire design

It is important to remember when designing a questionnaire that the questionnaires have to be analysed. If 50 questionnaires are returned, there has to be a system of analysing them. When you design a questionnaire, you should have an idea of the kind of answers you are likely to receive.

Here are some rules to remember when designing a questionnaire.

1 Keep the questionnaire short (one or two pages).
2 Start with very simple questions.
3 Give as many codified response questions as possible (questions with a range of answers to choose from).
4 Keep open-ended or complicated questions to a minimum. Better still, have none at all.
5 Keep complicated or delicate questions to the end (name, address, telephone number, salary, etc. are all delicate questions).

EXAMPLE 5

(a) Start with very simple questions (see Fig 11.2).

| Do you still own the car that was recently serviced by us? | Yes ☐ |
| | No ☐ |

Fig 11.2

(b) Give as many codified response questions as possible (see Fig 11.3).

(c) Keep open-ended or complicated questions to a minimum (see Fig 11.4).

(d) Keep complicated or delicate questions to the end (see Fig 11.5).

Please indicate by ticking the appropriate box how much you agree or disagree with the following statements.

	Agree strongly	Agree slightly	Neither agree nor disagree	Disagree slightly	Disagree strongly
I was attended to within a reasonable length of time	☐	☐	☐	☐	☐
The staff were courteous and helpful at all times	☐	☐	☐	☐	☐
The mechanic understood my requirements	☐	☐	☐	☐	☐
My car was left in safe and competent hands	☐	☐	☐	☐	☐
All my instructions were carried out satisfactorily	☐	☐	☐	☐	☐
The work was carried out to a high standard	☐	☐	☐	☐	☐
It is likely that I shall return for my next service	☐	☐	☐	☐	☐

Fig 11.3

Which aspect of service would you like to see improved?

_____

Fig 11.4

What are your views on euthanasia?

_____

Do you know why the House of Lords voted against euthanasia?

_____

Fig 11.5

Use of questionnaire

Who will we give the questionnaire to? This is a very good question. It is of no use to give a questionnaire intended for office workers to shop assistants. It is also very time consuming and costly to give a questionnaire to everyone in a large organisation.

■ Population

Population is a term used by statisticians. It refers to any group of items from which a sample may be taken. Some examples of population are:

- the employees of a certain company;
- the subscribers to a certain magazine;
- boxes of muesli on a production line;
- dresses made in a factory.

■ Sampling frame

The sampling frame is a complete list or complete identification of the population. Some examples of sampling frame are:

- The names and departments of employees in a certain company. (This information can be obtained from personnel records.)
- The names and addresses of subscribers to a certain magazine. (This information can be obtained from customer records.)
- Every box of muesli coming off a production line in any one day.
- Every dress made in a factory in any one day.

■ Sampling techniques

Sampling is the final stage used to identify exactly whom the questionnaire should be given to. There are five common sampling techniques:

- Simple random sampling
- Stratified sampling
- Systematic sampling
- Multi-stage sampling
- Quota sampling.

Simple random sampling

This technique is often know as the 'lottery method'. It is so named because it follows this procedure:

1 All the items from the sample frame are given a number.
2 The numbers are placed in a hat or box.

3 A proportion of numbers is chosen at random.

4 The chosen numbers indicate who should be given the questionnaire to complete.

This procedure is acceptable if you are only dealing with a small sample frame. A larger sample frame will require the use of a computer-generated random table.

Stratified sampling

The word stratified means to form into layers or status groups. A stratified sample is therefore one that divides the sample frame into groups. Some examples of stratified groups are:

- Full-time and part-time workers
- Male and female
- Under 18 years and over 18 years
- Cars with four doors and cars with less than four doors
- Biscuits with cream, biscuits with chocolate and biscuits with cream and chocolate
- Accounts department, personnel department, marketing department, administration department, etc.

This technique can be described in three steps:

1 The sample frame is divided into the required groups.

2 The proportion of each group to the whole is calculated (e.g. 120 male workers and 80 female workers in an organisation will be 60% male and 40% female).

3 A simple random sample is taken from each group. (This is also calculated by proportion, e.g. of the 60% male workers 6 may be questioned and of the 40% female workers 4 may be questioned.)

Systematic sampling

This technique is one of the easiest and least expensive of all the techniques. From the sample frame every nth item is chosen. Some examples of systematic sampling are:

- If the sample frame is a list of 100 workers and 10% are to be questioned every 10th worker is chosen.
- If the sample frame is the telephone directory, open any page and choose every 20th person.

- From a production line every 50th toy is checked for faults. (The disadvantage of using this technique for a production line is that the machinery could be faulty at regular intervals.)

Multi-stage sampling

Multi-stage sampling is mainly used when the sample frame is very large or geographically spread out: for example, if the Department of Transport wished to survey drivers' views. The sample frame would be every adult with a current driving licence. To random sample this population could mean interviewers going to many parts of the country. This would be a very costly and time consuming exercise. Multi-stage sampling will reduce the costs and time required.

The procedure the technique follows is:

1 The country is divided into a few large regions. (Independent Television regions are a popular choice of market research companies.)
2 Two or three of these large regions are chosen at random.
3 The two or three chosen regions are subdivided into smaller areas. (Parliamentary constituencies or local authority regions are a common choice.)
4 A few of these smaller areas are chosen at random.
5 From a map of the few chosen smaller areas streets are chosen at random.
6 House numbers from each chosen street are random sampled until a list of individuals is obtained.

Figure 11.6 illustrates the technique.

Quota sampling

This technique will not give a great deal of accuracy. However, it is popular with market research companies as it is easy to implement. The 'quota' is a set number of a particular group from the sample frame. Some examples of quota samples are:

- 30 women aged 18–30
- 20 American males
- 30 unemployed men and 20 unemployed women aged 16–20
- 100 white-collar males aged 30–65
- 20 grandmothers.

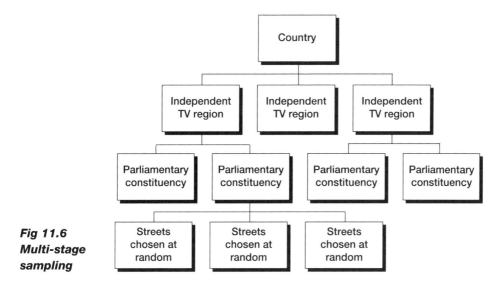

**Fig 11.6
Multi-stage
sampling**

Collecting data by observation

Observation is a valid and common method of data collection. Some data can only be collected by observation: for example, how many men and how many women attend a certain theatre performance. The total number of people in attendance can be taken from the tickets sold. However, the breakdown of men and women can only be carried out by observation (unless customers buy male and female tickets). Much observation can be performed by computer now; although the trained observer is still required in many industries.

Observation is a record of what happened. For example, how long it took a team of five men to build a brick wall 12 feet high and 40 feet long. It does not record whether the men thought the task to be strenuous or difficult, but it may record how many bricks and how much mortar they used.

Observation can be used to establish whether working patterns or communication methods in an organisation are adequate. For this kind of observation the people being observed must remain unaware of the observer. This is therefore a disadvantage as the observer must be trained to remain inconspicuous. Another disadvantage of observation is that it tells us nothing about people's attitudes (this can however be established in a questionnaire). Observation also gives us no indication about past or future data.

12 Recording data

Discrete and continuous data

Data is classified as either *discrete* or *continuous*. Discrete data is collected information which can only have certain values, such as the number of employees in an organisation. This data can only take whole number values like 0, 5, 100, 3000, etc. It cannot be $5\frac{1}{2}$, $10\frac{1}{2}$, etc. Continuous data is collected information which can take any value between two given values, such as the weights of aeroplane passengers. This data can be 70 kg, 75.4 kg or 78.2573 kg, depending upon the accuracy of measurement.

■ Raw data

Raw data is collected information which has not been arranged or organised in any way.

EXAMPLE 1

Table 12.1 shows the number of children in the family for 30 employees of a certain company.

Table 12.1

2	0	1	3	2	0	3	2	0	5
1	2	3	2	3	1	4	3	2	1
2	4	4	3	0	2	1	5	2	3

Frequency distributions

A frequency distribution is one way of organising raw data. The frequency distribution can take the form of an ungrouped frequency distribution or a grouped frequency distribution.

■ Ungrouped frequency distribution

Ungrouped frequency distributions are used when there are few categories of data, such as the number of children in a family.

EXAMPLE 2

Table 12.2 shows the data from Table 12.1 arranged as an ungrouped frequency distribution.

Table 12.2

Number of children per employee	Frequency
0	4
1	5
2	9
3	7
4	3
5	2

■ Grouped frequency distribution

Grouped frequency distributions are used when there are a lot of categories, such as the salaries of 50 employees. Employees may earn £15 000, £15 125, £25 370, etc. We would not wish to take this data and arrange it in increments of £1, therefore we group it together. Each group is called the class width. The size of the class width will depend upon the data collected and a sensible assessment of the collected information.

EXAMPLE 3

Table 12.3 shows the grouped frequency distribution of the salaries of 50 employees in a company.

Table 12.3

Salary of employees	*Frequency*
Under £9999	7
£10 000 to £14 999	13
£15 000 to £19 999	17
£20 000 to £24 999	9
£25 000 plus	4

Tally charts

A tally chart is the ideal method for recording data and then creating a frequency distribution.

EXAMPLE 4

A care assistant observes the stock of cereal in a children's home.

Cereal	Tally	Frequency
Cornflakes	卌 III	8
Shreddies	卌 卌	10
Rice Crispies	III	3
Sugar Puffs	卌 卌 II	12
Muesli	卌IIII	9
Ready Brek	卌 卌 卌	15

Fig 12.1 A tally chart

A vertical line is given for each packet of cereal. Four lines are grouped with a fifth horizontal line. This is the tally. Frequency is the tally converted into whole numbers.

A two-way table

A two-way table is another method of recording data.

EXAMPLE 5

(a) The output of five assembly workers was observed over a week and recorded on a two-way table (see Fig 12.2).

Worker \ Day	Mon	Tue	Wed	Thur	Fri
M Jones	157	155	157	150	160
K Fitz	170	169	166	170	168
J Clark	165	164	164	164	164
S Oliver	147	148	148	148	147
D Simpson	150	150	151	150	149

Fig 12.2 A two-way table

(b) 10 nurses were questioned about hospital administration systems. The nurses were asked their views on the efficiency of the administration systems. The results were displayed on a two-way table (see Fig 12.3).

Administration system \ Nurse	1	2	3	4	5	6	7	8	9	10
Patients records	VG	G	VG	B	VG	G	A	G	VG	A
Medication stock	VG	G	G	B	VG	G	A	VG	VG	A
Equipment stock	VG	G	A	B	B	G	G	VG	VG	A
Doctors roster	G	G	A	G	A	G	G	VG	G	B
Nurses roster	G	A	A	VB	VG	G	A	B	VG	A

VG = Very Good ; G = Good ; A = Average ; B = Bad ; VB = Very Bad

Fig 12.3

Exercise 12.1

1 A car park attendant observed the length of stay of 30 cars to the nearest 15 minutes. Produce a tally chart and frequency distribution for this information.

15	30	60	15	75	45	75	120	30	90
30	120	75	90	135	45	75	105	45	60
60	75	120	45	75	60	90	75	60	120

2 Design a two-way table for a supervisor of five labourers. The supervisor wishes to observe the time it takes each labourer to accomplish the following activities:

- Lay a course of bricks 10 m long
- Mix a metre of concrete
- Dig two fence post holes
- Erect 5 m of scaffolding.

3 Design and use a tally chart to observe the types and number of each type of fire extinguisher in a building.

4 Design and use a two-way table to observe the prices of five different products in three competing stores.

13 Converting between different units of measurement

In Chapter 7 it was mentioned that there are people who still use imperial units of measurement. Because both forms are still in use it is important that you can convert between the two. Conversion tables, graphs and scales will help you do this, but you should also know the more common measures in your head.

You will find it helpful to learn the following:

$$2.54 \text{ cm} = 1 \text{ inch}$$
$$1 \text{ metre} = 39.37 \text{ inches}$$
$$1 \text{ kilometre} = 0.6214 \text{ miles}$$
$$28.35 \text{ grams} = 1 \text{ oz}$$
$$1 \text{ kilogram} = 2.205 \text{ lb}$$
$$1 \text{ litre} = 0.22 \text{ gallons}$$

Conversion tables

Conversion tables such as those that appear in Table 13.1 (on page 128) are extremely useful for quick reference.

Table 13.1

Inches		Millimtrs	Feet		Metres	Miles		Kilomtrs	Yards		Metres
0.039	1	25.400	3.281	1	0.305	0.621	1	1.609	1.094	1	0.914
0.079	2	50.800	6.562	2	0.610	1.243	2	3.219	2.187	2	1.829
0.118	3	76.200	9.842	3	0.914	1.864	3	4.828	3.281	3	2.743
0.158	4	101.600	13.123	4	1.219	2.486	4	6.437	4.375	4	3.658
0.197	5	127.000	16.404	5	1.524	3.107	5	8.047	5.468	5	4.572
0.236	6	152.400	19.685	6	1.829	3.728	6	9.656	6.562	6	5.486
0.276	7	177.800	22.966	7	2.134	4.350	7	11.265	7.655	7	6.401
0.315	8	203.200	26.246	8	2.438	4.971	8	12.875	8.749	8	7.315
0.354	9	228.600	29.527	9	2.743	5.592	9	14.484	9.843	9	8.230

Ounces		Grams	Pounds		Kilogrms	Gallons		Litres	Pints		Litres
0.035	1	28.350	2.205	1	0.454	0.220	1	4.546	1.760	1	0.568
0.071	2	56.699	4.409	2	0.907	0.440	2	9.092	3.520	2	1.137
0.106	3	85.049	6.614	3	1.361	0.660	3	13.638	5.279	3	1.705
0.141	4	113.398	8.819	4	1.814	0.880	4	18.184	7.039	4	2.273
0.176	5	141.748	11.023	5	2.268	1.100	5	22.730	8.799	5	2.841
0.212	6	170.097	13.228	6	2.722	1.320	6	27.277	10.559	6	3.410
0.247	7	198.447	15.432	7	3.175	1.540	7	31.823	12.318	7	3.978
0.282	8	226.796	17.637	8	3.629	1.760	8	36.369	14.078	8	4.546
0.318	9	255.146	18.842	9	4.082	1.980	9	40.915	15.838	9	5.114

The numbers in bold in the centre of each table can indicate either measure.

EXAMPLE 1

(a) Convert 1 ounce into grams.

Ounces		Grams
0.035	1 →	28.350
0.071	2	56.699

Read from left to right.
Ounces → 1 → 28.350 g

(b) Convert 2 grams into ounces.

Ounces			Grams
0.035	**1**		28.350
0.071	← **2**		56.699

Read from right to left.

Grams   → **2** →   0.071 oz

Exercise 13.1

1 Convert the following:

   (a) 8 feet into metres
   (b) 4 kilograms into pounds
   (c) 5 litres into gallons

2 A filing cabinet weighs 70 kilograms. How much is this in pounds?

3 A canister holds 52 litres of water. How much is this in gallons?

4 My pen is 4 inches long. How many millimetres is this?

5 The distance from Redcar to Guisborough is 9 miles. How far is this in kilometres?

Conversion formulas

Sometimes it is easier to work with conversion formula tables (see Table 13.2 on page 130).

Table 13.2

To convert	Multiply by	To convert	Multiply by
Inches to centimetres	2.54	Acres to hectares	0.4047
Centimetres to inches	0.3937	Hectares to acres	2.471
Feet to metres	0.3048	Gallons to litres	4.546
Metres to feet	3.281	Litres to gallons	0.22
Yards to metres	0.9144	Ounces to grams	28.35
Metres to yards	1.094	Grams to ounces	0.03527
Miles to kilometres	1.609	Pounds to kilograms	0.4536
Kilometres to miles	0.6214	Kilograms to pounds	2.205

EXAMPLE 2

Convert 100 litres to gallons using Table 13.2.

From the right side of the table we read that to convert litres to gallons you multiply by 0.22.

$$100 \times 0.22 = 22 \text{ gallons}$$

Exercise 13.2

1 A petrol tank holds 10 gallons. How many litres will it hold?

2 If a journey is 48 miles, how many kilometres is it?

3 A farmer has 40 acres of land. What is this in hectares?

4 A block of margarine weighs 500 grams. How much does it weigh in ounces?

5 An elderly lady wishes to buy a washing machine 60 cm wide. She does not know if this will fit into her kitchen as she has measured the gap in inches. Convert 60 cm to inches.

Conversion scales

Figures 13.1 to 13.5 show the conversion scales of some of the most common units of measurement.

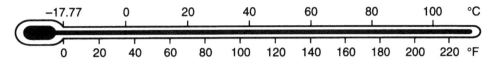

Fig 13.1 Conversion scale for °C to°F

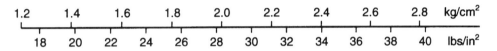

Fig 13.2 Conversion scale for kg/cm² to lbs/in²

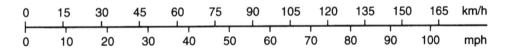

Fig 13.3 Conversion scale for km/h to mph

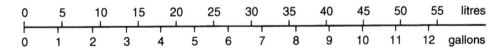

Fig 13.4 Conversion scale for litres to gallons

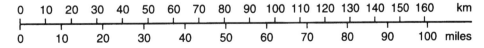

Fig 13.5 Conversion scale for km to miles

EXAMPLE 3

Convert 40 mph to km/h.

Read along the bottom line of Fig 13.3 to 40 mph.

The answer is 64 km/h.

Remember, scales are not as accurate as tables. Sometimes your answer will have to be your best educated guess.

Exercise 13.3

Using the conversion scales, answer the following questions.

1 A car travels on the motorway at 112 km/h. What is its speed in mph?

2 The pressure of each tyre on a car should be 24 lb/in^2. What is this in kg/cm^2?

3 An elderly woman has a body temperature of 95 °F. What is this in °C?

4 A tank holds 12 gallons of water. What is this in litres?

5 It is 70 miles from Leeds to Derby. What is this in km?

Conversion graphs

Figures 13.6 to 13.9 show the conversion graphs for some common units of measurement.

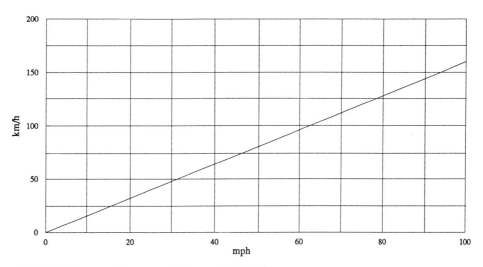

Fig 13.6 Conversion graph for mph to km/h

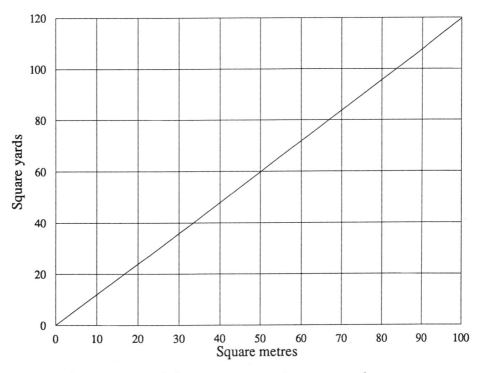

Fig 13.7 Conversion graph for square metres to square yards

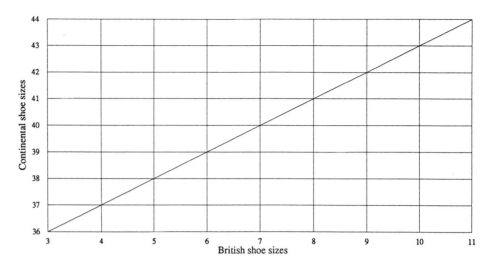

Fig 13.8 Conversion graph for British and Continental shoe sizes

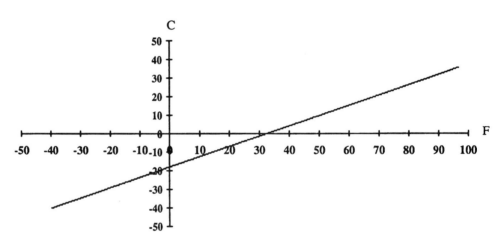

Fig 13.9 Conversion graph for Fahrenheit to Centigrade

EXAMPLE 4

Using Fig 13.6, convert 60 mph into km/h.

Use a ruler to draw in a vertical line from 60 on the x axis to the graph line. Draw a horizontal line from the graph line to the y axis. Read the value off the y axis.

Answer is 97 km/h.

Exercise 13.4

1 Using Fig 13.8, work out what the Continental size for a British shoe size 6 is.

2 Using Fig 13.7, convert 45 square metres into square yards.

3 Using Fig 13.9, express 50 °F in °C.

4 A salesman has a choice between two company cars. Car A accelerates from 0 to 60 mph in 9 seconds. Car B accelerates from 0 to 100 km/h in 9 seconds. Which car is faster?

5 A children's home requires new carpet in the day room. The room measures 80 yds^2. The showroom has two carpets that are ideal. Carpet A is £5 per m^2 and carpet B is a remnant of 70 m^2 costing £340. Which is the cheaper carpet?

6 A friend from the USA boasted that he sat through an American Football game when the temperature was – 14°F. During the Winter Olympics you sat through temperatures reaching – 20°C. Which is colder?

14 Averages and range

The purpose of averages

Information which has been collected via observation, a questionnaire or from statistical data may not always be easy to compare. The method of averages is used to make the data more 'user-friendly'. You may want to compare information from more than one survey or you may want to compare your survey results with government statistics, for example.

When making comparisons it is useful to have a single value that is typical of the distribution of data. This single value is known as the average. There are three types of averages: the *mean*, *median* and *mode*. You are probably used to using the mean as an average rather than the median or the mode, but sometimes they are more appropriate. It is important you choose the most appropriate average for the particular comparison you wish to make.

■ Mean

The *mean* of a set of raw data is found by adding all the observations together and dividing by the number of observations. The mean of a frequency distribution, however, is found by multiplying each observation (x) with its frequency (f) to get (fx), adding up the new values ($\sum$ fx) and dividing by the total frequency ($\sum$ f). When you see the $\sum$ symbol in front of a letter it means you have to total that column or row.

Raw data formula is: $\dfrac{\sum x}{n}$

Frequency distribution formula is: $\frac{\Sigma fx}{\Sigma f}$

EXAMPLE 1

(a) The mean of raw data

Ten light bulbs were tested to see how long they would burn for. The results, in hours, are:

200, 203, 214, 198, 212, 198, 220, 199, 199, 201

Find the mean $\left(\frac{\Sigma x}{n}\right)$:

$$\frac{200 + 203 + 214 + 198 + 212 + 198 + 220 + 199 + 199 + 201}{10}$$

$$= \frac{2040}{10} = 204$$

Answer = 204 hours.

(b) The mean of a frequency distribution

Table 14.1 (on page 138) shows the number of children in the family for 30 employees of a certain company.

Column x shows the number of children an employee may have.

Column f shows the number of employees with that many children (e.g. there are 3 employees with 4 children each).

Column fx shows the total number of children between the employees.

To obtain column fx you must multiply each line at a time; for example, line 1 shows that there are 4 employees with no children, therefore $0 \times 4 = 0$; line 2 shows that there are 5 employees with 1 child, therefore $1 \times 5 = 5$; etc.

To obtain the sum of the frequency column (Σf) you must add up the f column. This is the total number of employees.

To obtain the sum of the fx column (Σfx) you must add up the fx column. This is the total number of children.

Table 14.1

Number of children per employee	Frequency	Sum of children
x	f	fx
0	4	0
1	5	5
2	9	18
3	7	21
4	3	12
5	2	10
Sum	30	66

Substitute the totals into the formula:

$$\Sigma fx = 66$$

$$\Sigma f = 30$$

$$\frac{\Sigma fx}{\Sigma f} = \frac{66}{30} = 2\tfrac{1}{5} \text{ or } 2.2$$

Answer = the average number of children per family is 2.2 children.

■ Median

The *median* of a set of data is the middle value. You must first arrange all the data into size order (smallest number to largest number) and then find the middle value. If there are 11 observations then the 6th value is the median. This is found by adding 1 to 11 = 12, then dividing by 2. If there are 10 observations then $1 + 10 = 11$; $11 \div 2 = 5\tfrac{1}{2}$, so the $5\tfrac{1}{2}$th value is the median.

EXAMPLE 2

(a) **The median of raw data**

    (i) Find the median of the following set of numbers:

       8 5 9 6 5 7 7 5 4

First arrange in size order:

4 5 5 5 6 7 7 8 9

There are 9 values therefore 9 + 1 = 10; 10 ÷ 2 = 5th value

Answer = 6 is the median number.

(ii) Find the median of the following set of numbers:

22 25 21 25 27 25 26 21 23 22 21 26

First arrange in size order:

21 21 21 22 22 23 25 25 25 26 26 27

There are 12 values, therefore 12 + 1 = 13; 13 ÷ 2 = $6\frac{1}{2}$th value

Because the median falls between the 6th and the 7th values you must add the 6th and 7th values together then divide by 2:

23 + 25 = 48; 48 ÷ 2 = 24

Answer = 24 is the median number.

(b) **The median of a frequency distribution**

Table 14.2 shows the number of visits 30 young mothers made to the family doctor over a three month period. Find the median number of visits made to the doctor.

Table 14.2

Number of visits x	Frequency f
4	9
5	5
6	7
7	4
8	3
9	2

The (x) column tells you the number of visits any one mother made to the doctor and the (f) column tells you how many mothers made those number of visits: for example (f) 9 and (x) 4 means 9 mothers made only 4 visits each to the doctor in the three month period.

To find the median number of visits to the doctor you must first find the median mother. There are 30 mothers therefore $30 \div 2 = $ 15th mother (we do not add 1 in frequency distributions).

To find the 15th mother you must change the table into a cumulative frequency or 'less than' table (see Table 14.3).

Table 14.3

Number of visits	Frequency	Number of visits 'less than'	Cumulative frequency	
x	f	x		cf
		less than 4 visits	0	0
4	9	< 5	0+9=9	9
5	5	< 6	0+9+5=14	14
6	7	< 7	0+9+5+7=21	21
7	4	< 8	0+9+5+7+4=25	25
8	3	< 9	0+9+5+7+4+3=28	28
9	2	< 10	0+9+5+7+4+3+2=30	30

We can see from the cumulative frequency that the 14th mother pays less than 6 visits to the doctor and the 21st mother pays less than 7 visits to the doctor, so the median mother pays less than 7 visits to the doctor.

Answer = the median number of visits to the doctor is 6.

■ Mode

The *mode* of a set of data is the value which occurs most often. The advantage of the mode is that its value can occur, whereas with the mean or median the value is a calculation, for example 2.2 children as in Example 1.

EXAMPLE 3

(a) **The mode of raw data**

(i) Find the mode of the following set of numbers:

7 5 6 9 8 3 7 7 2 4 6 7 3

Because 7 occurs 4 times, this is the mode.

Answer = the mode is 7.

(ii) Find the mode of the following set of numbers:

22 24 25 29 19 20 21 27 23

Answer = there is no mode.

(iii) Find the mode of the following set of numbers:

12 13 15 19 16 12 14 17 15 12 16 15

Answer= the mode is 12 and 15.

This is known as bi-modal.

(b) The mode of a frequency distribution

Table 14.4 shows the number of children in the family for 30 employees of a certain company. Find the mode number of children per family.

The mode of a frequency distribution is simply the highest frequency.

Table 14.4

Number of children per employee	Frequency
x	f
0	4
1	5
2	9
3	7
4	3
5	2

The highest frequency is 9 therefore the mode number of children per family is 2.

Answer = the mode number of children per family is 2 children.

Exercise 14.1

1 Find the mean of the following:

    (a) 5 8 9 6 5 4 6 3 7 8 6 5

    (b) 33 36 35 34 35 32 36 38 34 37

    (c) 1.7 1.4 1.3 1.5 1.6 1.5 1.4 1.6

2 Find the median of the following:

    (a) 75 78 82 42 43 59 64 89 67 62 57

    (b) 106 105 104 102 103 105 108 107 106

    (c) 3.01 3.2 3.05 2.01 3.75 2.5 2.85

3 Find the mode of the following:

    (a) 56 58 59 54 57 56 54 54 59 53 56 57 54 52 54

    (b) 1132 1134 1135 1132 1136 1132 1135

    (c) 0.002 0.003 0.006 0.005 0.002 0.005 0.004 0.001 0.005

4 Below are the number of birthday cards a shop assistant sold each day for 14 days. Find the mean, median and mode number of birthday cards sold.

127 132 112 145 127 138 115 113 127 141 150 124 136 119

5 Below are the weekly wages of 20 builders. Find the mean, median and mode weekly wage.

£120 £125 £110 £130 £120 £105 £110 £115 £120 £135
£130 £115 £120 £110 £115 £120 £130 £100 £105 £100

6 Calculate the mean, median and mode of the following frequency distribution:

(a)

x	1	2	3	4	5	6	7
f	5	8	7	10	12	3	3

(b)

x	547	548	549	550	551	552	553	554
f	11	16	22	26	25	20	15	15

7 100 boxes of screws were sampled and the contents counted. The results of the survey are shown in Table 14.5. Find the mean, median and mode number of screws per box.

Table 14.5

Number of screws per box x	Frequency f
197	3
198	9
199	16
200	45
201	20
202	5
203	2

Range

The range of a set of data is the difference between the smallest value and the largest value. It is found by subtracting the smallest value from the largest value. The range on its own is not a very useful measure, but used alongside the mean it can highlight unusual data. The range identifies extreme values.

EXAMPLE 4

(a) Ten light bulbs were tested to see how long they would burn for. The results, in hours, are:

200, 203, 214, 198, 212, 198, 220, 199, 199, 201

Find the mean and the range.

The range is: $220 - 198 = 22$

In Example 1 we found the mean life of the light bulbs to be 204 hours.

By calculating both the mean and the range we can decide how close to the original data the average is. If the range is large we know that somewhere in the set of data is an extreme value. In this example the extreme value is 220 hours. If we exclude this value from the calculation of range and mean the results are:

Mean   = 202.6 hours
Range   = 16 hours

(b) Compare the performance of two salesmen by using the mean and range of their weekly orders (see Table 14.6).

Table 14.6

Week	1	2	3	4	5	6
Salesman A	33	86	50	91	56	98
Salesman B	74	59	80	52	71	48

Salesman A: Mean = 69   **Salesman B:**   Mean = 64
                Range = 65   Range = 32

Salesman A has a higher mean (69 orders) than Salesman B (64 orders). However, B has shown more consistency in his work by having a smaller range.

Exercise 14.2

1 Calculate the mean and range for the following:

     (a)   10, 9, 11, 8, 13, 7, 14, 16
     (b)   114, 87, 131, 75, 151, 66, 173, 57, 195, 50, 44, 57
     (c)   98, 95, 92, 94, 91, 99 ,96
     (d)   251, 281, 225, 249, 258, 272, 256, 262, 258, 268

2 Calculate the mean and range for the monthly sales figures for Product A and Product B in 199–, as shown in Table 14.7.

Table 14.7

Month	Sales of A (£)	Sales of B (£)
January	2000	1000
February	2800	1500
March	3040	2500
April	3560	3440
May	4600	5020
June	5080	7800
July	5700	8500
August	6800	6000
September	7580	5400
October	7600	4500
November	7800	3000
December	8000	2200

If the company were to cease to produce one of the products, which would you recommend and why?

3 Compare the performance of two machines that pack sugar. Ten sample packets are weighed, correct to the nearest gram, and their weights recorded below. Calculate the mean and range for each machine. What recommendations would you make to the company regarding the two machines?

Machine A   495, 496, 496, 498, 499, 500, 500, 501, 501, 504

Machine B   492, 494, 494, 499, 500, 501, 503, 504, 506, 507

Quartiles

The quartiles of a set of data divide the data into four even sections (quarters). The median is the middle quartile, known as Q_2. The other quartiles are Q_1 and Q_3. To find Q_1 and Q_3 you must follow a similar procedure as for finding the median.

EXAMPLE 5

A sample of 11 weekly wages from a certain company were taken and are shown below. Find the quartile weekly wages.

£375 £2000 £410 £220 £600 £470 £140 £260 £335 £300 £200

A simple formula helps to find which observation represents each quartile.

$$Q_1 = \tfrac{1}{4}(n + 1)$$

$$Q_2 = \tfrac{1}{2}(n + 1)$$

$$Q_3 = \tfrac{3}{4}(n + 1)$$

where n = number of observations.

In this example the number of observations is 11, therefore

$$Q_1 = \tfrac{1}{4}(11 + 1) = \text{3rd observation}$$

$$Q_2 = \tfrac{1}{2}(11 + 1) = \text{6th observation}$$

$$Q_3 = \tfrac{3}{4}(11 + 1) = \text{9th observation}$$

First arrange the data in size order:

£140 £200 £220 £260 £300 £335 £375 £410 £470 £600 £2000

$Q_1 = £220$

$Q_2 = £335$

$Q_3 = £470$

■ Interquartile range

The interquartile range of a set of data is the range of the middle 50% of the data, from Q_1 to Q_3. Let us use the sample of 11 weekly wages from Example 5. You can see from the data that there is a weekly wage which is much larger than the other wages. This is the wage of the managing director of the company. If we were to include this value into a calculation of the range the data would show a range of £1860. The interquartile range will however exclude this extreme value. The interquartile range calculates as £250. This is a much more realistic figure. When using the full range of data a mean of £482.73 is obtained, but when using the interquartile range we get a mean of only £338.57. This value is more realistic.

There are times when data is being analysed that one should exclude certain extreme results because they distort some calculations. There are other times when extreme results should be included because they give the true picture.

Exercise 14. 3

1 Calculate the quartiles and the interquartile range of the following:

(a) 7 5 2 9 7 6 4 5 3 7 3
(b) 22 65 35 34 31 28 57 54 25 26 68 49 21 26 42
(c) 8 5 7 6 3 8 2 4 9 6 8 2 1 8 7 3 5 4 7

2 Below are the values of a computer sales representative's last 11 sales. Calculate the quartiles and the interquartile range of the sales figures.

| £22 000 | £50 500 | £46 000 | £12 000 | £500 | £4000 |
| £1500 | £5750 | £19 120 | £375 | £162 000 | |

15 Statistical diagrams

Statistical diagrams are essentially concerned with displaying data to the greatest advantage. A sales team would be keen to have management and customers see its impressive increase in sales. A line graph, like the one in Fig 15.1, with a broad red line indicating sales figures, will give an immediately favourable representation of this information.

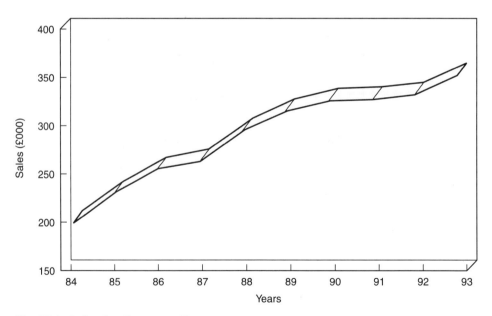

Fig 15.1 Sales for Company X

At this point it is necessary to point out that data is either discrete or continuous. You must know the difference. *Discrete data* is any data that has exact values. For example:

- the number of children in a family
- the number of light bulbs made in a factory
- the number of goals scored in a football match
- the shoe sizes of members of a health club.

Continuous data is any data that can only be measured to a certain degree of accuracy. It cannot assume any exact value. For example:

- the heights of children in a family correct to the nearest cm
- the life of 20 light bulbs to the nearest minute
- the time taken to make a widget to the nearest minute
- the speed of cars passing a checkpoint.

Pictograms

A *pictogram* is the simplest form of displaying data. A picture is selected that easily identifies the data, such as a large £ sign to represent salaries or a man to represent male employees. One picture is used to represent a specified value of the data (e.g. £1000) and the picture is then repeated for every multiple of that value (e.g. 5 £ signs = £5000). The picture (or symbol) should never be represented in a larger form for a larger quantity. Example 1 shows the correct way and the incorrect way to create a pictogram.

EXAMPLE 1

A company sells 50 000 boxes of soap detergent per week. Display this information in the form of a pictogram (see Fig 15.2 on page 150).

The advantage of a pictogram is that it is very easy to understand. Its disadvantages are that it is not very accurate and it can be difficult to construct.

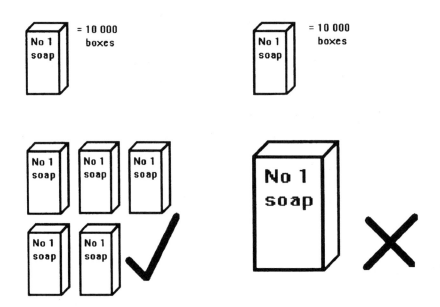

Fig 15.2 The right and wrong way to create a pictogram

Reminders on construction of graphs

1 The information on the y axis always depends upon the information on the x axis. For example, the monthly sales figures depend upon the month of the year. Therefore, sales figures are on the y axis and months on the x axis. The population of major British cities depends upon the city. Therefore, population is on the y axis and the cities are on the x axis.

| y axis | vertical axis | dependent variable |
| x axis | horizontal axis | independent variable |

2 Choose a scale for each axis. Some common scales are:

1 cm = 1 unit	1 cm = 5 units	1 cm = 10 units
1 cm = 25 units	1 cm = 100 units	1 cm = 250 units
1 cm = 1000 units	1 cm = 2000 units	

3 When displaying data on a line graph you join the points with a ruler. You only join points freehand if you are using a curve to calculate gradient, etc.

4 You do not join any of the points on a scatter graph.

5 Give a main title to the graph and titles to the x and y axes.

EXAMPLE 2

Figure 15.3 shows the differences between a line graph and a bar chart. Both graphs are displaying the same information.

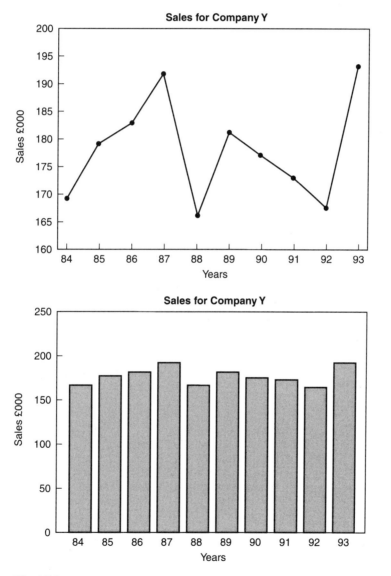

Fig 15.3

Exercise 15.1

1 The number of orders received by a manufacturing company over ten weeks was as shown in Table 15.1.

Table 15.1

Week	No. of orders
1	20
2	44
3	56
4	82
5	120
6	110
7	96
8	68
9	50
10	12

Construct a line graph for the data.

2 B & D's production of crude steel, 1989–94, is shown in Table 15.2.

Table 15.2

Year	Production (000 metric tons)
1989	174
1990	189
1991	187
1992	178
1993	164
1994	164

Construct a bar chart for the data in the table.

3 The monthly sales figures for Product A and Product B in 199– are shown in Table 15.3.

Table 15.3

Month	Sales of A (£)	Sales of B (£)
January	2000	1000
February	2800	1500
March	3040	2500
April	3560	3440
May	4600	5020
June	5080	7800
July	5700	8500
August	6800	6000
September	7580	5400
October	7600	4500
November	7800	3000
December	8000	2200

Construct both a bar chart and a line graph for the data in the table.

4 A company has 8000 male employees and 9500 female employees. Display this information in the form of a pictogram.

Scatter graphs

A scatter graph is used to show a correlation or relationship between two variables. It is a XY graph where the X axis is the independent variable and the Y axis the dependent variable. Each point is plotted on the graph, but the points are not joined up.

The main problem with constructing scatter graphs is deciding which variable goes on which axis. Remember the rule:

y axis	vertical axis	dependent variable
x axis	horizontal axis	independent variable

EXAMPLE 3

A factory worker was observed over a week. Fig 15.4 on page 154 shows the number of items made during each hour. The information was then displayed as a scatter graph (see Fig 15.5 on page 154).

Day \ Hours	10	11	12	2	3	4	5
Monday	150	148	147	140	130	120	100
Tuesday	148	147	147	140	135	130	128
Wednesday	155	150	140	135	125	120	120
Thursday	145	145	140	135	130	130	110
Friday	130	128	127	120	110	108	100

Fig 15.4 Chart for number of items made

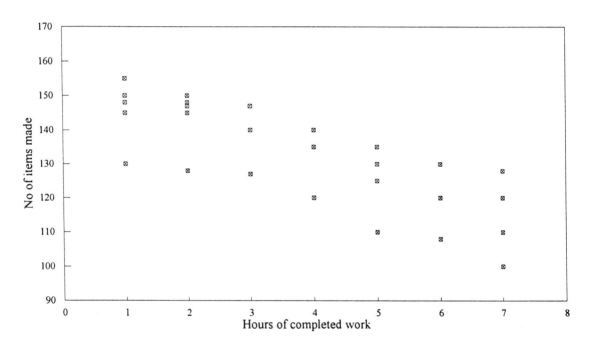

Fig 15.5 Scatter graph showing number of items made during each hour

The scatter graph in Fig 15.5 shows that the worker was able to make more items at the beginning of the day and less at the end of the day. Could this possibly be because of tiredness in the afternoon?

Scatter graphs can be any one of three different types: negative correlation, positive correlation and no correlation. The scatter graph in Example 2 is a negative correlation. The more time the worker spends at work, the less work he does.

Figure 15.6 shows the three different types of scatter diagrams.

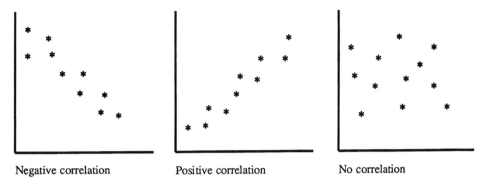

Negative correlation   Positive correlation   No correlation

Fig 15.6 The three types of scatter graph

EXAMPLE 4

Figure 15.7 shows three scatter graphs.

In Fig 15.7 (a) there is no correlation between marketing expenditure and sales. This means that any increase in marketing expenditure has not necessarily had a positive effect on sales.

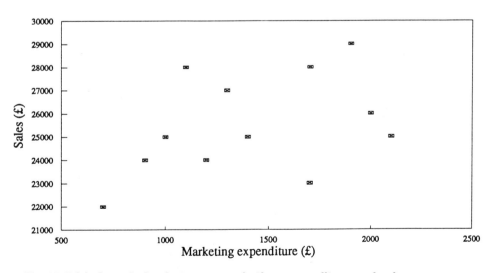

Fig 15.7 (a) Correlation between marketing expenditure and sales

In Fig 15.7 (b) there is a positive correlation between ice-cream sales and temperature. This means that as the temperature increased, the sale of ice-cream increased.

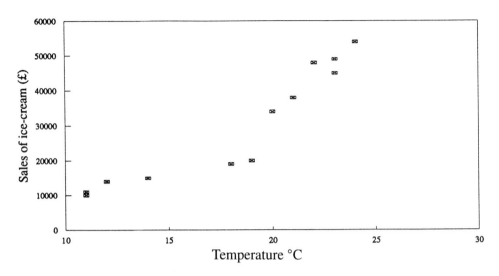

Fig 15.7 (b) Correlation between ice-cream sales and temperature

In Fig 15.7 (c) there is a negative correlation between demand and price. This means that there is less demand for expensive items and greater demand for cheaper items.

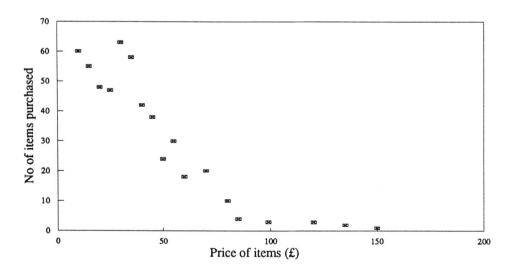

Fig 15.7 (c) Correlation between demand and price

Exercise 15.2

Construct a scatter graph of the information presented in Table 15.4.

Table 15.4

Number of hours of personal study	Grade for assignments
13	A
8	C
3	E
7	B
4	D
6	D
8	C
9	D
11	B
10	B
12	A

A is the higher grade and E is the lower grade.

Reminders on construction of pie charts

- There are 360° in a circle.
- Indicate the number of degrees or the percentage of each section somewhere in your work.
- Give your pie chart a title, and name each slice.

EXAMPLE 5

The advertising costs for Brits & Co. in 199– were as shown in Table 15.5.

Table 15.5

Advertising method	Cost (£000)
Exhibitions	70
Newspapers	65
Posters	25
Radio	45
TV	95

Add up all the costs.

Answer = 300.

The number of degrees for each section of pie are:

Exhibitions $\dfrac{70}{300} \times 360 = 84°$

Newspapers $\dfrac{65}{300} \times 360 = 78°$

Posters $\dfrac{25}{300} \times 360 = 30°$

Radio $\dfrac{45}{300} \times 360 = 54°$

TV $\dfrac{95}{300} \times 360 = 114°$

The right-hand column should always add up to 360°. The data is presented in the pie chart in Fig 15.8.

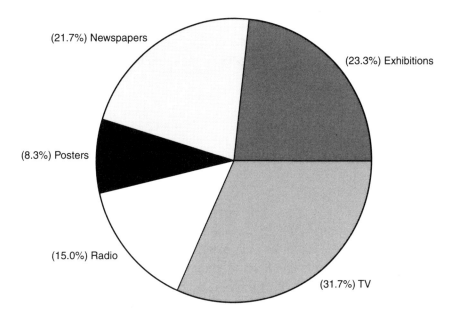

Fig 15.8 Advertising costs for Brits & Co. 199–

Exercise 15.3

1 The Wendel Nursing Home has 50 residents. Their ages are grouped together and the distribution is shown in Table 15.6.

Table 15.6

Age	Number of residents
60–64	5
65–69	9
70–74	13
75–79	15
80–85	6
85 +	2

Construct a pie chart for the data.

2 A factory manufactures two products X and Y. Construct two pie charts to compare costs (see Table 15.7 on page 160).

Table 15.7

Cost	Product X (£000)	Product Y (£000)
Direct labour	58	48
Direct materials	30	65
Overheads	27	12
Administration	5	15

Histograms

A histogram is a statistical diagram which represents a frequency distribution. It looks very similar to a bar chart, but is not constructed in the same way. The *area* of each bar represents the frequency of the various classes. If all the classes on a grouped frequency distribution are the same width then all the bars will be of equal width, and the height of the bars will be determined by the frequency. But if the classes are of different widths then the bars will also be of different sizes and the heights calculated accordingly.

EXAMPLE 6

(a) Table 15.8 shows a frequency distribution of the time a sample of employees spend talking to customers on the telephone. Construct a histogram of the data.

Table 15.8

Number of seconds talking on the telephone	Frequency
1–30	25
31–60	43
61–90	128
91–120	246
121–150	175
151–180	130
181–210	78
211–240	64
241–270	51
271–300	33

Before you can construct the histogram you must first identify the lower and upper class boundaries. For the class 1–30 seconds the lower class boundary is 0.5 seconds and the upper class boundary is 30.5 seconds. For the class 31–60 seconds the lower class boundary is 30.5 seconds and the upper class boundary is 60.5 seconds, etc.

Figure 15.9 shows the histogram of the data.

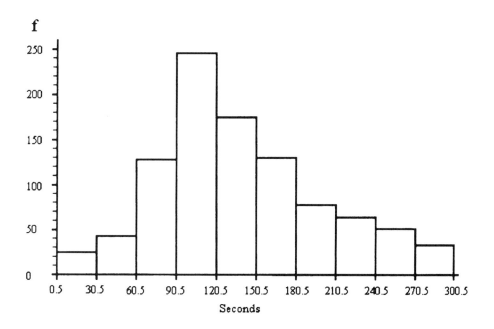

Fig 15.9 Histogram to show time spent talking to customers on the telephone

Interpret statistical diagrams

Statistical diagrams contain most of their information in the body of the graph or chart. The main title and axes titles also contain information. When you are asked to interpret statistical diagrams write your answer in full (usually one or two sentences).

Exercise 15.4

Average number of doctors' visits per year
by children and young people

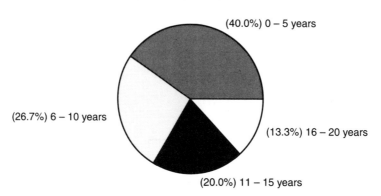

Fig 15.10 (a)

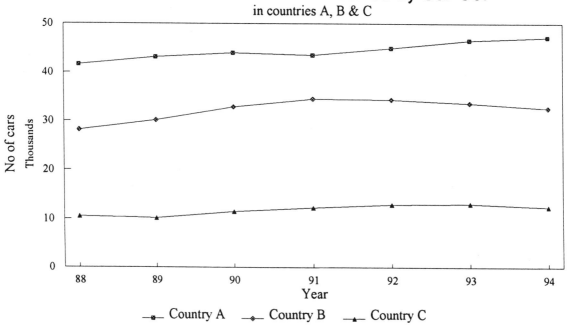

Fig 15.10 (b)

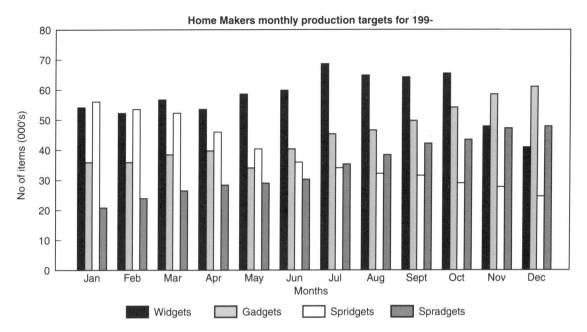

Fig 15.10 (c)

To answer these questions you must first identify which of the diagrams in Fig 15.10 gives you the required information.

1 What year had the highest number of cars manufactured by Car Co. in any one country?

2 What age range had the highest number of visits to the doctor?

3 What country manufactured less cars in 1991 than they did in 1990?

4 What is the highest number of any one item produced in one month by Home Makers? In what month does this happen?

5 Which of Home Makers' products increases in production over the year?

16 Equations and inequalities

Simple equations

Simple equations are usually linear equations. A linear equation, when drawn on a graph, will always produce a straight line. There is an important rule to remember when solving linear equations:

- What you do to one side you must do to the other. (For example, if you add 5 to the right-hand side (RHS), you must add 5 to the left-hand side (LHS).) Alternatively you may have learned it as:
- Change sides = change signs. (For example, + changes to − ; × changes to ÷, and vice versa.)

EXAMPLE 1

(a) Solve $5x + 7 = 27$

$5x + 7 - 7 = 27 - 7$ (here 7 has been subtracted from both sides)

$5x = 20$

$\dfrac{5x}{5} = \dfrac{20}{5}$ (here 5 has been divided into both sides)

or alternatively:

Solve $5x + 7 = 27$

$5x = 27 - 7$ (here + 7 changed sides and became − 7)

$$5x = 20$$

$$5 = \frac{20}{5} \quad \text{(here} \times 5 \text{ changed sides and became} \div 5\text{)}$$

$$x = 4$$

(b) A theatre has 150 seats and the revenue for a single performance is £1680. Assuming every seat was sold, what is the cost of each seat?

Let the cost of each seat be x.
Therefore

$$150x = £1680$$

$$\frac{150x}{150} = \frac{£1680}{150} \quad \text{(here both sides are divided by 150)}$$

$$x = £11.20$$

or alternatively:

$$150x = £1680$$

$$x = \frac{£1680}{150} \quad \text{(here} \times 150 \text{ changed sides and became} \div 150\text{)}$$

$$x = £11.20$$

(c) A painter spends £80.75 and buys 6 small canvases and 5 large canvases. The large canvases cost £3.50 more than the small canvases. What is the cost of a small canvas?

Let the cost of the small canvas be x.
Therefore the cost of the large canvas will be $x + £3.50$.

$$6x + 5(x + 3.5) = 80.75 \quad \text{(you must first multiply out the brackets)}$$

$$6x + 5x + 17.5 = 80.75$$

$$11x + 17.5 - 17.5 = 80.75 - 17.5$$

$$11x = 63.25$$

$$\frac{11x}{11} = \frac{63.25}{11}$$

$$x = £5.75$$

Exercise 16.1

1 Solve the following equations:

(a) $x + 8 = 15$

(b) $3x + 7 = 22$

(c) $120x - 40 = 200$

(d) $75x - 425 = 175$

(e) $\frac{x}{5} = 12$

2 Five chairs cost £549.95. What is the cost of each chair?

3 It takes 12 hours to make 4 objects. How long does it take to make 1 object?

4 A salesman sells 350 objects and earns £962.50. How much does he earn per object?

5 A crew row 24 miles in 3 hours. How many miles per hour do they row?

6 Workers receive a basic wage of £35 plus £1.27 per item. If a worker makes 158 items in a week, what will be the gross wage?

7 A construction site buys 10 bags of cement and 60 bags of sand. The cement costs three times as much as the sand. If the total bill is £90, what is the cost of a bag of cement and the cost of a bag of sand?

8 A computer software salesman sells a database package and a spreadsheet package. The database package costs £21.50 more than the spreadsheet package. In a particular week he sold 15 database packages and 21 spreadsheet packages with a total revenue of £1632 . What is the cost of the spreadsheet package?

9 A community centre spends £500 on 30 chairs. They buy soft chairs and hard chairs. The hard chairs cost £15 each and the soft chairs cost £20 each. How many of the hard chairs do they buy?

10 A sports centre rents its hall out at £42.50 per hour. The hall can be divided into two basketball courts and 2 badminton courts or 1 basketball court and 6 badminton courts or 10 badminton courts. Assuming the costs are proportional to size, what is the hourly cost of (a) a badminton court and (b) a basketball court?

Simultaneous equations and inequalities

Simultaneous equations are two equations that are connected. The x values in the two equations are identical. The y values in the equations follow the same rule.

Inequalities are equations where one term is greater than or less than the other term. You need to remember and know the meaning of the following symbols:

$>$ greater than $\qquad\qquad$ $<$ less than

$\geqslant$ greater than or equal to $\qquad$ $\leqslant$ less than or equal to

Simultaneous equations and inequalities are needed in industry because it is not always possible to expect things to run smoothly. For example, two machines may be used to manufacture two different products. Only one product can be manufactured at a time on each machine. Simultaneous equations are used to help solve the problem of how many of each product will give the best usage of machinery.

Inequalities are needed because it is not always possible to have a perfect solution. For example, you may have 5 workers in a factory available for 8 hours in a particular day. This does not mean that you have 40 hours of production time. One worker may not be able to use a machine another worker uses. Machines or workers may be idle while waiting for others to finish.

■ Solving simultaneous equations

EXAMPLE 2

(a) Solve the simultaneous equations:

$$x - 3y = 17 \qquad (1)$$
$$4x - 3y = 8 \qquad (2)$$

Add Equations (1) and (2):

$$(x + 4x) + (3y - 3y) = 17 + 8$$
$$5x + 0 = 25$$

$$5x = 25$$
$$x = \frac{25}{5}$$
$$x = 5$$

Substitute 5 into Equation (1):

$$5 + 3y = 17$$
$$3y = 17 - 5$$
$$3y = 12$$
$$y = \frac{12}{3}$$
$$y = 4$$

Check in Equation (2):

$$(4 \times 5) - (3 \times 4) = 8$$
$$20 - 12 = 8$$

(b) Solve the simultaneous equations:

$$2x + 6y = 20 \qquad (1)$$
$$3x + 2y = 16 \qquad (2)$$

Multiply Equation (1) by 3:

$$6x + 18y = 60 \qquad (3) \qquad \text{(this becomes Equation (3))}$$

Multiply Equation (2) by 2:

$$6x + 4y = 32 \qquad (4) \qquad \text{(this becomes Equation (4))}$$

Subtract Equation (4) from Equation (3):

$$(6x - 6x) + (18y - 4y) = 60 - 32$$
$$0 + 14y = 28$$
$$14y = 28$$
$$y = \frac{28}{14}$$
$$y = 2$$

Substitute 2 into Equation (1):

$$2x + 12 = 20$$
$$2x = 20 - 12$$
$$2x = 8$$
$$x = \frac{8}{2}$$
$$x = 4$$

Check in Equation (2):

$$(3 \times 4) + (2 \times 2) = 16$$
$$12 + 4 = 16$$

Exercise 16.2

Solve the following simultaneous equations for x and y.

1 $x + 5y = 13$

$6x - 5y = 8$

2 $2x + 6y = 32$

$4x + 3y = 37$

3 $4x + y = 48$

$3x - 3y = 6$

4 $3x + 2y = 79$

$2x + 5y = 115$

5 $7x - 5y = 8$

$2x + 7y = 36$

■ Solving simultaneous equation problems

EXAMPLE 3

A theatre has 300 seats and charges adults £18 and children £15. At one performance the total revenue was £5160. How many adults and how many children attended the performance?

Let the number of adults be x and the number of children be y.

$$x + y = 300 \quad (1)$$
$$18x + 15y = 5160 \quad (2)$$

Multiply Equation (1) by 15:

$$15x + 15y = 4500 \quad (3)$$

Subtract Equation (3) from Equation (2):

$$(18x - 15x) + (15y - 15y) = 5160 - 4500$$
$$3x + 0 = 660$$
$$3x = 660$$
$$x = \frac{660}{3}$$
$$x = 220$$

Substitute 220 in Equation (1):

$$220 + y = 300$$
$$y = 300 - 220$$
$$y = 80$$

Answer = 220 adults and 80 children were at the performance.

Exercise 16.3

1 A bill for £195 was paid with £5 notes and £10 notes. If 24 notes are used, how many £5 notes were there?

2 Two machines, A and B, are used to make product x and product y. Product x needs 1 hour on Machine A and 3 hours on Machine B. Product y needs 2 hours on Machine A and 2 hours on Machine B. Machine A can only be used for 16 hours continuously but Machine B can be used for 36 hours. Calculate the number of each product that should be produced to utilise the maximum capacity of the machines.

3 A factory has 1100 m of a certain cloth out of which it makes dresses and shirts. A dress uses 4 m of the cloth and a shirt uses 2 m. The dress sells at £30 and the shirt sells at £22. If total revenue is £10 000, how many of each item are made?

4 Peter works 38 hours basic and 4 hours overtime and earns £352. Matthew works 40 hours basic and 6 hours overtime and earns £392. Find x and y if x is the basic pay per hour and y is the overtime pay per hour.

■ Solve simple inequalities

Here are some examples of inequalities:

$x > 3$ means an unknown value x is greater than 3 (but not including 3)

$x \geq 3$ means an unknown value x is greater than or equal to 3 (including 3)

The rules for solving inequalities are the same as for solving equations (e.g. change sides = change sign) with one exception: the inequality sign must be reversed when multiplying or dividing both sides by a negative number.

EXAMPLE 4

Solve the inequality:

$$4 - 5x < 20 - 3x$$

Move x values to left-hand side of inequality:

$$-5x + 3x < 20 - 4$$
$$-2x < 16$$
$$x > \frac{16}{-2}$$
$$x > -8$$

Exercise 16.4

Solve the following inequalities:

1 $5 + 4x > 9 + 2x$

2 $11 + 6x \geqslant 18 - x$

3 $22 - 2x \geqslant 7 + 3x$

4 $19 + 3x \geqslant 2x - 5$

5 $25 + 2x < 8x - 9$

6 A nurse has two activities to complete in less than 120 minutes. Activity 1 is to roll bandages and Activity 2 is to make up pressure pads. The bandages take 2 minutes to roll and the pressure pads take 5 minutes to make. Write an expression for this problem. If she has to have 14 pressure pads, how many bandages can she make?

7 A factory produces widgets (x) and gadgets (y). It takes one hour and 3 kg of raw materials to make a widget. It takes two hours and 1 kg of raw materials to make a gadget. The factory can only produce one item at a time and has only 64 hours per week production time. The factory is also constrained in that it is only delivered 72 kg of raw materials per week. How many of each product should it make to maximise production? Write an expression for the time constraint and an expression for the raw material constraint. Solve the equations.

Answers to exercises

EXERCISE 1.2

1 $\frac{1}{5}$

2 $\frac{1}{10}$

3 $\frac{1}{9}$

4 $\frac{1}{2}$

5 $\frac{1}{3}$

6 25%

7 25%

8 40%

9 10%

10 2%

EXERCISE 1.3

1 No. It is an 11-hour working day with probably no breaks.

2 No. 1000 squeegee mops will take 10 months to sell.

3 Yes. It will take 2 hours 5 minutes to type.

4 No. The data does not support the argument.

5 Yes. 8 pages per min × 5 mins = 40 pages.

EXERCISE 1.4

1 £3

2 6 m

3 34°C

4 25 l

5 489 km

6 £140

7 380 m

8 170 kg/m^2

9 1270 km

10 590 books

11 27 500 miles

12 £135 200

13 13 700 bricks

14 8600 leaflets

15 62 600 kb of disk space free

EXERCISE 1.5

1 6 m

2 £80

3 70 kg

4 73 000 kb of disk space

5 £380 000

EXERCISE 1.6

1 Between £6960 to £10 560

2 Between £1360 to £1480

3 Between 24 615 litres and 30 968 litres

EXERCISE 2.1

1 (a) 21
   (b) 24
   (c) 32
   (d) 50
   (e) 90
   (f) 13
   (g) 26
   (h) 12
   (i) 34
   (j) 8
   (k) 8
   (l) 12
   (m) 36
   (n) 9
   (0) 9

2 £1100

3 348

4 (a) £387.75
   (b) £537.94
   (c) £1444.78
   (d) £665.00
   (e) £3035.47

5 132 907

6 (a) Machinist D makes 355 kites
   (b) 1594 kites

7
Portable cassette player	12
Portable CD player	19
Mini hi-fi system	8
Midi hi-fi system	9
Car radio	4
Television	7
Video player	8
Satellite system	3
Washing machine	6
Tumble dryer	14
Refrigerator	6
Freezer	4
Vacuum cleaner	17

8 443.75 m

9 0.127 mm

10 12 521 291 components

EXERCISE 2.2

1 (a) 54
   (b) 24
   (c) 28
   (d) 99
   (e) 84
   (f) 48
   (g) 63
   (h) 36
   (i) 42
   (j) 72
   (k) 4
   (l) 4
   (m) 3
   (n) 11
   (0) 7

2 £28 144.13

3 (a) 9 lengths
(b) 44 lengths
(c) £203.01

4 (a) 448 bricks
(b) £143.36

5 $8 \times [(250 \times £28.75) + (450 \times £22.25) + (800 \times £18.50)] = 8 \times £3200 = £256\,000$

EXERCISE 2.3

1 −101

2 −83 151

3 −£826

4 £6221

5 460 bags

EXERCISE 2.4

1 (a) 8×10^8
(b) 2.7×10^7
(c) 4.9×10^6
(d) 1.23×10^{11}
(e) 8×10^{12}

2 (a) 400 000
(b) 32 000 000
(c) 910 000 000 000
(d) 7 900 000 000
(e) 1 100 000 000 000 000

3 3×10^8

4 1.12×10^9

5 1.5×10^{11}

EXERCISE 3.1

1 (a) $\frac{7}{2}$
(b) $\frac{23}{5}$
(c) $\frac{21}{4}$
(d) $\frac{30}{11}$
(e) $\frac{25}{7}$

2 (a) $2\frac{3}{4}$
(b) $7\frac{1}{3}$
(c) $2\frac{3}{8}$
(d) $3\frac{2}{5}$
(e) $3\frac{5}{6}$

3 (a) $\frac{1}{4}$
(b) $\frac{1}{2}$
(c) $\frac{7}{9}$
(d) $\frac{11}{16}$
(e) $\frac{17}{22}$

4 (a) $\frac{4}{5}$
(b) $\frac{14}{21}$ and $\frac{6}{21}$ $Answer = \frac{2}{3}$
(c) $\frac{5}{40}$ and $\frac{32}{40}$ $Answer = \frac{4}{5}$
(d) $\frac{77}{99}$ and $\frac{45}{99}$ $Answer = \frac{7}{9}$
(e) $\frac{150}{325}$ and $\frac{143}{325}$ $Answer = \frac{6}{13}$

5 (a) $\frac{8}{12} + \frac{3}{12} = \frac{11}{12}$
(b) $\frac{3}{6} + \frac{5}{6} = \frac{8}{6}$
(c) $\frac{15}{20} + \frac{16}{20} = \frac{31}{20} = 1\frac{11}{20}$
(d) $\frac{10}{30} + \frac{12}{30} + \frac{15}{30} = \frac{37}{30} = 1\frac{7}{30}$
(e) $\frac{6}{24} + \frac{4}{24} + \frac{16}{24} = \frac{26}{24} = 1\frac{2}{24} = 1\frac{1}{12}$

6 (a) $\frac{3}{8} - \frac{2}{8} = \frac{1}{8}$

(b) $\frac{11}{16} - \frac{10}{16} = \frac{1}{16}$

(c) $\frac{21}{35} - \frac{5}{35} = \frac{16}{35}$

(d) $\frac{15}{18} - \frac{8}{18} = \frac{7}{18}$

(e) $\frac{10}{16} - \frac{7}{16} = \frac{3}{16}$

7 (a) $\frac{4}{26} = \frac{2}{13}$

(b) $\frac{16}{63}$

(c) $\frac{4}{17} \times \frac{2}{1} = \frac{8}{17}$

(d) $\frac{1}{20} \times \frac{25}{3} = \frac{1}{4} \times \frac{5}{3} = \frac{5}{12}$

(e) $\frac{1}{22} \times \frac{11}{2} = \frac{1}{2} \times \frac{1}{2} = \frac{1}{4}$

8 (a) $\frac{3}{8} \times \frac{8}{5} = \frac{3}{5}$

(b) $\frac{7}{16} \times \frac{4}{1} = \frac{7}{4} = 1\frac{3}{4}$

(c) $\frac{13}{21} \times \frac{7}{3} = \frac{13}{3} \times \frac{1}{3} = \frac{13}{9} = 1\frac{4}{9}$

(d) $\frac{15}{18} \times \frac{12}{13} = \frac{15}{3} \times \frac{2}{13} = \frac{30}{39} = \frac{10}{13}$

(e) $\frac{4}{9} \times \frac{27}{24} = \frac{1}{1} \times \frac{3}{6} = \frac{3}{6} = \frac{1}{2}$

EXERCISE 3.2

1 (a) 1.63
(b) 231.879
(c) 47.67
(d) 5074.72
(e) 819.734

2 (a) 377.8
(b) 21.1
(c) 521.419
(d) 42465.4
(e) 39353.502

3 (a) 14.28
(b) 2601.36
(c) 17.9228
(d) 0.008544

(e) 24381.666

4 (a) 1.32
(b) 880
(c) 54
(d) 4.56
(e) 63.57

EXERCISE 3.3

1 (a) 271
(b) 76
(c) 320 g
(d) 9045 m
(e) £174.30

2 (a) 546
(b) 66
(c) 720 kg
(d) 90 km
(e) 60p

3 (a) 20%
(b) 25%
(c) 12%
(d) 33%
(e) 44.5%

4 (a) 1000
(b) 7500
(c) £86
(d) 47 584 km
(e) 9000 m

EXERCISE 3.4

1 (a) Total parts $2 + 3 = 5$
Quantity per part $350 \div 5 = 70$
Multiply out
$$2 \times 70 : 3 \times 70 = 140 : 210$$
Answer = £140 : £210

(b) Total parts $\qquad$ $13 + 12 = 25$
Quantity per part $\qquad$ $1000 \div 25 = 400$
Multiply out
$\qquad 13 \times 40 : 12 \times 40 = 520 : 480$
Answer = 520 km : 480 km

(c) Total parts $\qquad$ $3 + 2 = 5$
Quantity per part $\qquad$ $155 \div 5 = 31$
Multiply out
$\qquad 3 \times 31 : 2 \times 31 = 93 : 62$
Answer = 93 g : 62 g

(d) Total parts $\qquad$ $19 + 11 = 30$
Quantity per part $\qquad$ $660 \div 30 = 22$
Multiply out
$\qquad 19 \times 22 : 11 \times 22 = 418 : 242$
Answer = 418 cm : 242 cm

(e) Total parts $\qquad$ $27 + 20 = 47$
Quantity per part $\qquad$ $282 \div 47 = 6$
Multiply out
$\qquad 27 \times 6 : 20 \times 6 = 162 : 120$
Answer = 162 ml : 120 ml

2 (a) Total parts $\qquad$ $2 + 3 + 2 = 7$
Quantity per part $\qquad$ $700 \div 7 = 100$
Multiply out
$\qquad 2 \times 100 : 3 \times 100 : 2 \times 100 = 200 : 300 : 200$
Answer = 200 g : 300 g : 200 g

(b) Total parts $\qquad$ $11 + 9 + 7 = 27$
Quantity per part $\qquad$ $243 \div 27 = 9$
Multiply out
$\qquad 11 \times 9 : 9 \times 9 : 7 \times 9 = 99 : 81 : 63$
Answer = 99 km : 81 km : 63 km

(c) Total parts $\qquad$ $8 + 5 + 3 = 16$
Quantity per part $\qquad$ $128 \div 16 = 8$
Multiply out
$\qquad 8 \times 8 : 5 \times 8 : 3 \times 8 = 64 : 40 : 24$
Answer = 64 litres : 40 litres : 24 litres

(d) Total parts $\qquad$ $2 + 1 + 3 = 6$
Quantity per part $\qquad$ $5946 \div 6 = 991$
Multiply out
$2 \times 991 : 1 \times 991 : 3 \times 991 = 1982 : 991 : 2973$
$\qquad$ Answer = £1982 : £991 : £2973

(e) Total parts $\qquad$ $17 + 9 + 4 = 30$
Quantity per part $\qquad$ $1.5 \div 30 = 0.05$
Multiply out
$17 \times 0.05 : 9 \times 0.05 : 4 \times 0.05 = 0.85 : 0.45 : 0.2$
$\qquad$ Answer = 0.85 m : 0.45 m : 0.2 m

3 (a) 3:5
(b) 4:5
(c) 3:6 and 4:6 Answer = 2:3
(d) 55:77 and 21:77 Answer = 5:7
(e) 54:675 and 75:675 Answer = 3:27

EXERCISE 3.5

1 Increase is £35 – £28 = £7
Percentage increase = $\frac{7}{28} \times 100 = 25$
Answer = 25%

2 $\frac{375}{600} = \frac{5}{8}$

3 Overspend is £10 700 – £10 000 = £700
Percentage overspend = $\frac{700}{10\,000} \times 100 = 7$
Answer = 7%

4 $\frac{375}{500} = \frac{75}{100} = \frac{3}{4} = 0.75$

5 (a) $\frac{12}{76} \times 100 = 15.79$
Answer = 15.79%
(b) $\frac{54}{76} \times 100 = 71.05$
Answer = 71.05%

6 200 : 500
Answer = 2:5

7 15 ml : 30 ml

Answer = 1:2

EXERCISE 3.6

1 $\frac{15}{100} \times 25 = 3.75$

£25 − £3.73 = £21.25

Answer = £21.25

2 $\frac{175}{1000} \times 1585 = 277.38$

£1585 + £277.38 = £1862.38

Answer = £1 862.38

3 Increase in output is

1 800 000 − 1 500 000 = 300 000 car parts

$\frac{300\,000}{1\,500\,000} = \frac{3}{15} = \frac{1}{5}$ therefore $\frac{1}{5} \times 100 = 20\%$

Answer = 20%

4 11:66 and 12:66

Answer = 2:11

5 $35 + 35 \times \frac{9}{100} = 38.15$

Answer = £38.15

6 $\frac{6}{835} x = 90$

$x = 90 \times \frac{835}{6} = 12\,525$

Answer = £12 525

7 2.5 cm represents 5 km

1 cm = 5 km ÷ 2.5 cm = 2 km

9 cm × 2 km = 18 km

Answer = 18 km

8 $365 \times \frac{12}{100} = 43.8$

Answer = £43.80

9 Total parts   13 + 11 + 8 = 32

Quantity per part   1000 ÷ 32 = 31.25

Multiply out

13 × 31.25 : 11 × 31.25 : 8 × 31.25 =

406.25 : 343.75: 250

Answer = £406.25 : £343.75 : £250

10 $500 \times \frac{7}{10} = 350$

11 Size of extension = Size of new building −
Size of old building

Size of new building = 37 500 m³ × 3.5
= 131 250 m³

Size of extension = 131 250 m³ − 37 500 m³
= 93 750 m³

12 Brand A costs £1.10 for 230 g

$\frac{115}{230}$ p for 1 g

0.5p for 1 g

Brand B costs £1.30 for 250g

$\frac{125}{250}$ p for 1 g

0.5p for 1 g

Both brands are the same.

13 $£\frac{24\,120}{2} = £12\,060$

$£\frac{24\,120}{3} = £8040$

$£\frac{24\,120}{8} = £3015$

14 Let total delivery = x

If rotten fruit is 8% then good fruit is 92%

92% of x kg = 552 kg

0.92 × x kg = 552 kg

x kg = 552 kg ÷ 0.92

x kg = 600 kg

EXERCISE 4.1

1 £10 000 + (£0.35 × 500) = £10 175

2 £505 600 − £17 850 = £487 750

3 564 pairs + 75 pairs = 639 pairs

4 £13.50 + (£.568 × 580) = £342.94

5 $\frac{2350}{27\,500}$ = £0.09

= 9 p

6 £980 – £205 – £55 = £720

7 10 182 litres –1350 litres = 8832 litres

EXERCISE 4.2

1 $\frac{176}{4}$ = 44 km/h

2 0.6 amperes × 5 ohms = 3 volts

3 £7000 × 0.08 × 3 = £1680

4 £2000 × 0.15 × 4 = £1200

5 $\left(\frac{9}{5} \times 50\right)$ + 32 = 122° F

6 $\frac{272}{20}$ = 13.6 g/cm^3

EXERCISE 5.1

1 (a) $3 \times 3 \times 3 = 27$
   (b) $2 \times 2 \times 2 \times 2 \times 2 = 32$
   (c) $10 \times 10 \times 10 \times 10 \times 10 \times 10 = 1\,000\,000$
   (d) $9 \times 9 = 8l$
   (e) $3 \times 3 \times 3 \times 3 = 81$

2 (a) 3
   (b) 4
   (c) 5
   (d) 4
   (e) 7

3 (a) 531 441
   (b) 78 125
   (c) 1 679 616
   (d) 571 623.75
   (e) 0.000000000134

4 (a) 0.12
   (b) 40
   (c) 9
   (d) 7
   (e) 3

EXERCISE 5.2

1 £3500(1 + 0.09)^2 = £4532.60

2 £1500(1 + 0.12)^4 = £2360.28

3 £15 500(1 + 0.055)^{12} = £29 468.72

4 700(1 + 0.11)^2 = 862.47
   862.47 – 700 = 162.47
   Interest = £678.13

5 950(1 + 0.08)^7 = 1628.13
   1628.13 – 950 = £678.13
   Interest = £678.13

EXERCISE 5.3

1 £5500(1 – 0.2)^5 = £1802.24

2 £15 000(1 – 0.12)^5 = £7915.98

3 £30 000 × (1 – 0.14)^{10} = £6639.05

4 300 = P(1 – 0.08)^5
   300 = P × 0.6590815
   455.18 = P
   Original value = £455.18

5 5000 = P(1 – 0.16)^5
   5000 = P × 0.4182119
   11 955.66 = P
   Original cost = £11 955.66

EXERCISE 6.1

1 $\frac{1}{2}$

2 (a) $\frac{1}{6}$

(b) $\frac{3}{6} = \frac{1}{2}$

(c) $\frac{2}{6} = \frac{1}{3}$

3 (a) $\frac{13}{25}$

(b) $\frac{12}{25}$

4 $\frac{4}{24} = \frac{1}{6}$

5 $\frac{15}{500} = \frac{3}{100}$

6 $\frac{6}{100} = \frac{3}{50}$

7 $\frac{280}{14\,000} = \frac{1}{50}$

8 $\frac{2}{250} = \frac{1}{125}$

9 $\frac{96}{300} = \frac{8}{25}$

10 $\frac{4}{125}$

EXERCISE 6.2

1 $\frac{1}{2} \times \frac{1}{2} = \frac{1}{4}$

2 (a) $\frac{1}{6} \times \frac{1}{6} = \frac{1}{36}$

(b) $\frac{3}{6} \times \frac{3}{6} = \frac{9}{36} = \frac{1}{4}$

(c) $\frac{2}{6} \times \frac{2}{6} = \frac{4}{36} = \frac{1}{9}$

3 (a) $\frac{13}{25} \times \frac{2}{5} = \frac{26}{125}$

(b) $\frac{12}{25} \times \frac{3}{5} = \frac{36}{125}$

4 $\frac{4}{24} \times \frac{2}{24} = \frac{8}{576} = \frac{1}{72}$

5 $\frac{280}{14\,000} \times \frac{250}{15\,000} = \frac{1}{50} \times \frac{1}{60} = \frac{1}{3000}$

EXERCISE 6.3

1 $\frac{2}{200} = \frac{1}{100} = 0.01$

2 $\frac{54}{450} = \frac{6}{50} = 0.12$

3 $\frac{54}{1000} = \frac{27}{500} = 0.054$

4 Product A: $\frac{349}{500} = 0.698$

Product B: $\frac{296}{428} = 0.692$

Product A is more efficient.

5 $\frac{2}{40} = \frac{1}{20} = 0.05$

EXERCISE 7.1

1 (a) 500 cm
(b) 3 cm
(c) 27 cm
(d) 427 cm

2 (a) 210 mm
(b) 3150 mm
(c) 46 mm
(d) 5000 mm

3 (a) 9 ft
(b) 4 ft
(c) 8 ft
(d) $2\frac{1}{2}$ ft

4 (a) 2 yd
(b) 4 yd
(c) 3520 yd
(d) 3 yd

5 (a) 5 m
   (b) 3000 m
   (c) 18 m
   (d) 4870 m

6 (a) 17 km
   (b) 5 km
   (c) 6.4 km
   (d) 64.3 km

7 (a) 2 g
   (b) 4000 g
   (c) 8.3 g
   (d) 17 540 g

8 (a) 7 kg
   (b) 25 kg
   (c) 3000 kg
   (d) 0.67 kg

9 (a) 32 oz
   (b) 672 oz
   (c) 71 oz
   (d) 352 oz

10 (a) 4 lb
   (b) 70 lb
   (c) 3 lb
   (d) 30 lb

11 (a) 2000 ml
   (b) 300 ml
   (c) 1450 ml
   (d) 740 ml

12 (a) 5 l
   (b) 2 l
   (c) 3.56 l
   (d) 4.9 l

13 (a) 2 pt
   (b) 64 pt
   (c) $2\frac{1}{2}$ pt
   (d) 33 pt

14 (a) 2 gal
   (b) 3 gal
   (c) $5\frac{1}{2}$ gal
   (d) $4\frac{1}{2}$ gal

EXERCISE 7.2

1 (a) 3 hr 50 min
   (b) 2 min
   (c) 19 hr 31 min
   (d) 3 days 2 hr

2 (a) 2 hr 30 min
   (b) 2 days 12 hr
   (c) 30 sec
   (d) 1 hr 18 min 17 sec

EXERCISE 7.3

1 45 litres ÷ 4.5 mins = 10 l/min

2 1 000 000 litres ÷ 330 mins = 3030.3 l/min

3 12 m ÷ 120 secs = 0.1 m/s

4 8 km ÷ 0.75 hour = 10.7 km/h

5 45 g ÷ 225 cm^3 = 0.2 g/cm^3

6 1200 customers ÷ 48 hours = 25 customers per hour

7 3000 words ÷ 500 ins^2 = 6 words per square inch

EXERCISE 8.1

1 (a) 5.2 m
   (b) 680 mm
   (c) 9 cm

(d) 37 m

(e) 27 m

2 2×5 m $+ 2 \times 3$ m $+ 2 \times 0.5$ m
$= 10$ m $+ 6$ m $+ 1$ m $= 17$ m

3 2×2206 m $+ 2 \times 3673$ m
$= 11\ 758$ m

4 2 circuits

5 1200 mm $+ 900$ mm $+ 900$ mm $= 3000$ mm

EXERCISE 8.2

(a) 16 cm^2

(b) 27 cm^2

(c) 12 cm^2

(d) 18 cm^2

(e) 22 cm^2

(f) 17 cm^2

EXERCISE 8.3

1 35 m^2

2 1856 cm$^2 = 0.1856$ m^2

3 $106\ 470$ m^2

4 $\frac{1}{2} \times 600$ mm $\times 1200$ mm $= 360\ 000$ mm^2
$= 3600$ cm^2
$= 0.36$ m^2

5 $\frac{1}{2} \times 60$ cm $\times 70$ cm $= 2100$ cm^2

6 45 m $\times 30$ m $= 1350$ m^2 (Yes)

7 Blood bank requires 6×5 m$^2 - 30$ m^2
Trellis Hall is 5 m $\times 9$ m $= 45$ m^2
Cumberland Hall is 7 m $\times 8$ m $= 56$ m^2

8 Parking bay of ferry is 25.2 m $\times 153$ m $=$ 3855.6 m^2

$\frac{3855.6}{16.2} = 238$

9 Chocolate boxes are 7.5 cm $\times 6.5$ cm $=$ 24.375 cm^2
195 cm^2 allocated by shop for x boxes of 24.375 cm^2

$x = \frac{195}{24.375} = 8$

10 Each sign is $\frac{1}{2} \times 0.6$ m $\times 0.6$ m $= 0.18$ m^2
Using a 2m^2 sheet of metal, x signs of 0.18 m^2 can be made.

$x = \frac{2}{0.18} = 11$

EXERCISE 8.4

1 20 mm $\times 20$ mm $\times 20$ mm $= 8000$ mm^3

2 3.5 m $\times 4.3$ m $\times 11.2$ m $= 168.56$ mm^3

3 47 cm $\times 62$ cm $\times 132$ cm $= 384\ 648$ cm^3
$= 0.385$ m^3

4 3000 mm $\times 2550$ mm$^2 = 7\ 650\ 000$ mm^3
$= 7650$ cm^3
$= 0.0077$ m^3

5 $\frac{1}{2} \times 30$ cm $\times 17$ cm $\times 5$ cm $= 1275$ cm^3

6 Old cabinets' volume $=$
5×50 cm $\times 60$ cm $\times 140$ cm $=$
$2\ 100\ 000$ cm^3
New cabinets' volume $=$
3×110 cm $\times 50$ cm $\times 200$ cm $=$
$3\ 300\ 000$ cm^3
Additional space $=$
$3\ 300\ 000$ cm$^3 - 2\ 100\ 000$ cm$^3 =$
$1\ 200\ 000$ cm^3

7 320×4.6 m $\times 3.2$ m $\times 8.7$ m $= 40\ 980.48$ m^3

8 Each charm has a volume of

$4 \text{ mm} \times 4.5 \text{ mm}^2 = 18 \text{ mm}^3$

There is 2430 mm³ available material to make x charms of 18 mm³.

$x = \frac{2430}{18} = 135$

9 30 cm × 32 cm × 57 cm contains 14 books with volume x.

54 720 cm³ contains 14 books with volume x.

$x = \frac{54720}{14} = 3908 \text{ cm}^3$

10 Volume of stone $= 60 \text{ cm} \times 60 \text{ cm} \times 90 \text{ cm}$
$= 324\,000 \text{ cm}^3$

Volume of support =

$\frac{1}{2} \times 60 \text{ cm} \times 60 \text{ cm} \times 90 \text{ cm} =$
162 000 cm³

Volume wasted $= 324\,000 \text{ cm}^3 - 162\,000 \text{ cm}^3$
$= 162\,000 \text{ cm}^3$

EXERCISE 8.5

1 $3.142 \times 30 \text{ mm} = 94.26 \text{ mm}$

2 $3.142 \times 5^2 = 78.54$ (78.55 cm²)

3 Circumference $= 3.142 \times 65 \text{ mm} \times 2$
$= 408.46 \text{ mm}$
Area $= 3.142 \times 65^2 \text{ mm}$
$= 13\,274.95 \text{ mm}^2$

4 $3.142 \times r^2 \quad = 50.272 \text{ m}^2$
$r^2 \quad = \frac{50.272}{3.142}$
$= 16 \text{ m}^2$
$r \quad = \sqrt{16 \text{ m}^2}$
$= 4 \text{ m}$

5 $3.142 \times d \quad = 9.426 \text{ cm}$
$d \quad = \frac{9.426}{3.142} \text{ cm}$
$= 3 \text{ cm}$

6 $3.142 \times \left(\frac{57}{2}\right)^2 = 2552.09$ (Answer 2552 cm²)

7 $3.142 \times 2.5 \text{ m} = 7.85 \text{ m}$

8 Circumference $= 3.142 \times 75 \text{ cm} \times 2$
$= 471.3 \text{ cm}$

Area $= 3.142 \times 75^2 \text{ cm}$
$= 17\,673 \text{ cm}^2$

9 $3.142 \times d \quad = 47\,130 \text{ mm}$
$d \quad = \frac{47\,130}{3.142} \text{ mm}$
$= 15\,000 \text{ mm}$

10 $3.142 \times r^2 \quad = 153 \text{ m}^2$
$r^2 \quad = \frac{153}{3.142} \text{ m}^2 = 49 \text{ m}^2$
$r \quad = \sqrt{49} \text{ m} = 7 \text{ m}$

EXERCISE 8.6

1 $3.142 \times 33^2 \text{ mm} \times 130 \text{ mm} = 444\,812.94 \text{ mm}^3$

2 Area of concrete $= 10 \text{ m} \times 25 \text{ m}$
$= 250 \text{ m}^2$

Area of grass $= (10 \text{ m} - 1.5 \text{ m} - 1.5 \text{ m}) \times$
$(25 \text{ m} - 1.5 \text{ m} - 1.5 \text{ m})$
$= 7 \text{ m} \times 22 \text{ m}$
$= 154 \text{ m}^2$

Area of pavement = Area of concrete –
Area of grass
$= 250 \text{ m}^2 - 154 \text{ m}^2$
$= 96 \text{ m}^2$

3 $3.142 \times \left(\frac{51}{2}\right)^2 - \left(\frac{48}{2}\right)^2 = 233.26$
Answer $= 233.26 \text{ mm}^2$

4 $3.142 \times \left(\frac{180}{2}\right)^2 \times 92 = 2\,341\,418.4 \text{ cm}^3$
Answer $= 2\,341\,418.4 \text{ cm}^3$

5 $(65 \times 24) + 2(65 \times 120) + 2(24 \times 120) =$
22 920
22 920 cm^2 = 2.29 m^2

6 Area of wood $\quad$ = 1.5 m^2
$\qquad\qquad\qquad\qquad$ = 15 000 cm^2
Area of table top $\quad$ = $3.142 \times \left(\frac{55}{2}\right)^2$ cm^2

$\qquad\qquad\qquad\qquad$ = 2376.14 cm^2
Area of support $\qquad$ = $2(50 \times 65)$ cm^2
$\qquad\qquad\qquad\qquad$ = 6500 cm^2
Area of waste =
Area of wood – Area of table top –
Area of support
$\qquad\qquad\quad$ = 6123.86 cm^2

7 Area of football pitch = 110 m × 75 m
$\qquad\qquad\qquad\qquad\quad$ = 8250 m^2
$\quad$ Area of roller = 3.142 × 1.2 m × 1.7 m
$\qquad\qquad\qquad$ = 6.41 m^2
No. of revolutions $\quad$ = Area of football pitch /
$\qquad\qquad\qquad\qquad\qquad$ Area of roller
$\qquad\qquad\qquad\qquad$ = 1287

8 Volume of Syringe A = 3.142×0.75^2 cm ×
$\qquad\qquad\qquad\qquad\qquad$ 8 cm
$\qquad\qquad\qquad\qquad$ = 14.13 cm^3
$\quad$ Volume of Syringe B = 3.142×0.625^2 cm ×
$\qquad\qquad\qquad\qquad\qquad$ 9 cm
$\qquad\qquad\qquad\qquad$ = 11.04 cm^3
Syringe A has the largest volume

EXERCISE 9.1

1 (a) cube
$\quad$ (b) cuboid
$\quad$ (c) triangular-based pyramid
$\quad$ (d) cylinder
$\quad$ (e) prism
$\quad$ (f) square-based pyramid

2 See Fig A1(a) – (e) (not to scale).

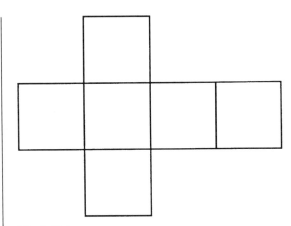

Fig A1(a)

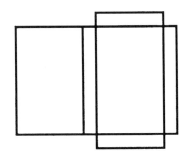

Fig A1(b)

Fig A1(c)

Fig A1(d)

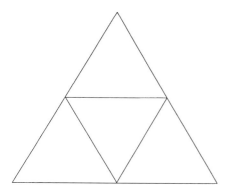

Fig A1(e)

EXERCISE 9.2

1 The objects are a 3-drawer filing cabinet, a can, a square-based pyramid and a carton (milk/juice).

2 See Fig A2(a) – (c).

Fig A2(b)

Fig A2(a)

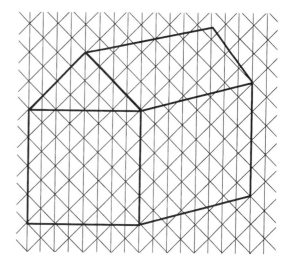

Fig A2(c)

EXERCISE 9.3

1 lounge

2 1

3 bathroom

4 kitchen

5 bedroom 2

EXERCISE 10.1

1 Birchet Road → St James Road → Gravel Street

2 See Fig A3.

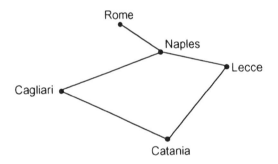

Fig A3

3 (a) London → Portsmouth → Cardiff → Birmingham → Liverpool → Manchester → Hull → Nottingham → London. See Fig A4.

Fig A4

(b) 750 miles

4 (a) See Fig A5.

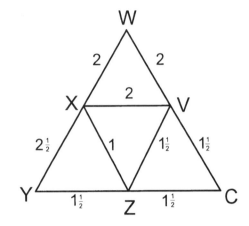

Fig A5

$$C \rightarrow V \rightarrow W \rightarrow X \rightarrow Y \rightarrow Z \rightarrow C$$

(b) 11 km

(c) $12\frac{1}{2}$ km

EXERCISE 10.2

1 See Fig A6.

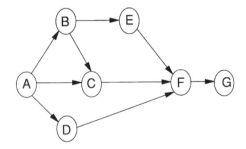

Fig A6

2 See Fig A7.

3 See Figs A8 and A9.

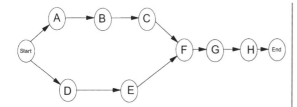

Fig A7

Activity	Prior Activity		Activity	Prior Activity
A	—		G	C
B	A		H	G
C	A		I	H
D	A		J	D,F,I
E	B		K	J
F	E		L	K

Fig A8

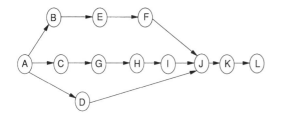

Fig A9

EXERCISE 12.1

1 See Fig A10.

Length of Stay	Tally	Frequency
15 minutes	II	2
30 minutes	III	3
45 minutes	IIII	4
60 minutes	⧻	5
75 minutes	⧻ II	7
90 minutes	III	3
105 minutes	I	1
120 minutes	IIII	4
135 minutes	I	1

Fig A10

2 See Fig A11.

Activity	Labourer	1	2	3	4	5
Lay a course of bricks 10 m long					ı	
Mix a metre of concrete						
Dig two fence post holes						
Erect 5 m of scaffolding						

Fig A11

EXERCISE 13.1

1 (a) 2.44 m
  (b) 8.82 lb
  (c) 1.1 gal

2 154.35 lb

3 11.44 gal

4 101.6 mm

5 14.48 km

EXERCISE 13.2

1 45.46 l

2 77.23 km

3 16.18 hectares

4 17.64 oz

5 23.62 in

EXERCISE 13.3

1 70 mph

2 1.7 kg/cm^2

3 35° C

4 54 l

5 112 km

EXERCISE 13.4

1 39

2 55 yd^2

3 9° C

4 Car B

5 Carpet A is cheaper

6 −14° F

EXERCISE 14.1

1 (a) $\frac{72}{12}$ = 6

   (b) $\frac{350}{10}$ = 35

   (c) $\frac{12}{8}$ = 1.5

2 (a) 64
   (b) 105
   (c) 3.01

3 (a) 54
   (b) 1132
   (c) 0.005

4 Mean = $\frac{1806}{14}$ = 129 cards

   Median = 127 cards
   Mode = 127 cards

5 Mean = $\frac{2335}{20}$ = £116.75

   Median = £117.50
   Mode = £120

6 (a) Mean = $\frac{181}{48}$ = 3.77

    Median = 4
    Mode = 5

   (b) Mean = $\frac{82\,583}{150}$ = 550.5

    Median = 550
    Mode = 550

7 Mean = $\frac{19\,993}{100}$ = 199.93 screws per box

   Median = 200 screws per box
   Mode = 200 screws per box

EXERCISE 14.2

1 (a) Mean = 11   Range = 9
   (b) Mean= 100   Range = 151
   (c) Mean = 95   Range = 8
   (d) Mean = 258   Range = 56

2

	Prod A	Prod B
Mean	5380	4233
Range	6000	7500

Product B because it has a lower average and a greater range.

	Machine A	Machine B
Mean	499	500
Range	9	15

Conclusions: Machine A is more consistent because it has a lower range, but its mean is lower than the required weight. Machine B has the required mean weight, but has a greater range and so is less consistent. Draw your own conclusions!

EXERCISE 14.3

1 (a) $Q_1 = 3$
$Q_2 = 5$
$Q_3 = 7$
Interquartile range = $7 - 3 = 4$

(b) $Q_1 = 26$
$Q_2 = 34$
$Q_3 = 54$
Interquartile range = $54 - 26 = 28$

(c) $Q_1 = 3$
$Q_2 = 6$
$Q_3 = 8$
Interquartile range = $8 - 3 = 5$

2 (c) $Q_1 = £1500$
$Q_2 = £12\,000$
$Q_3 = £46\,000$
Interquartile range = $£46\,000 - £1500$
$= £44\,500$

EXERCISE 15.1

1 See Fig A12.

Number of orders received in a 10 week period

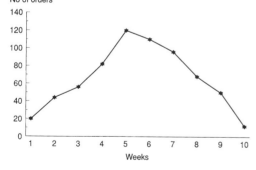

Fig A12

2 See Fig A13.

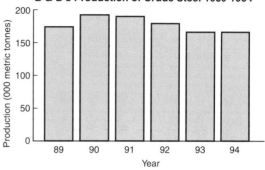

Fig A13

3 See Fig A14(a) and (b).

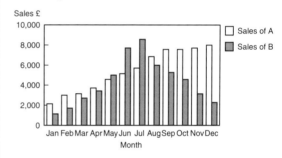

Fig A14(a)

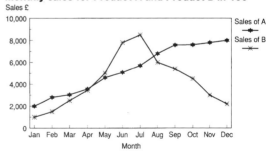

Fig A14(b)

4 See Fig A15.

Key

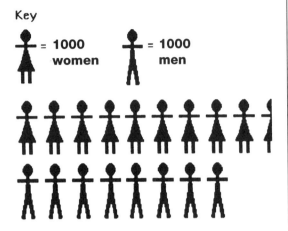

Fig A15

EXERCISE 15.2

See Fig A16.

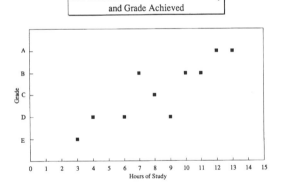

Fig A16

EXERCISE 15.3

1 See Fig A17.

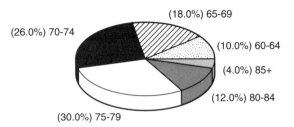

Fig A17

2 See Fig A18(a) and (b).

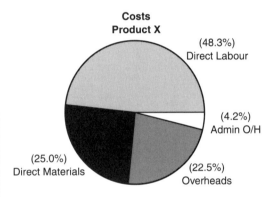

Fig A18(a)

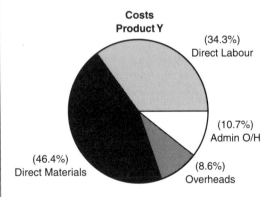

Fig A18(b)

EXERCISE 15.4

1 1994 had the highest number of cars manufactured by Car Co. It was in Country A.

2 0–5 year old children visited the doctor most often.

3 Country A manufactured less cars in 1991 than they did in 1990.

4 68 000 items was the highest number produced in any one month by Home Makers. This happened in July.

5 Spradgets increased in production over the year.

EXERCISE 16.1

1 (a) 7
(b) 5
(c) 2
(d) 8
(e) 60

2 £109.99

3 3 hours

4 £2.75

5 8 miles/hour

6 £35 + (£1.27 × 158) = £235.66

7 Let x = price of sand, therefore
$3x$ = price of cement
$$60x + (10 \times 3x) = 90$$
$$60x + 30x = 90$$
$$90x = 90$$
$$x = 1$$
Sand = £1 per bag, cement = £3 per bag

8 Let x = cost of the spreadsheet package, therefore $x + 21.5$ = cost of database package
$$15(x + 21.5) + 21x = 1632$$
$$15x + 322.5 + 21x = 1632$$
$$36x + 322.5 = 1632$$
$$36x = 1632 - 322.5$$
$$= 1309.5$$
$$x = 36.375$$
Spreadsheet package costs £36.38

9 Let x = number of hard chairs
$$15x + 20(30 - x) = 500$$
$$15x + 600 - 20x = 500$$
$$600 - 5x = 500$$
$$600 - 500 = 5x$$
$$100 = 5x$$
$$x = 20$$
They buy 20 hard chairs

10 (a) Let x = cost of the badminton court
$$10x = 42.5$$
$$x = 4.25$$
Hourly cost is £4.25
(b) Let y = cost of the basketball court, given that the badminton court costs £4.25, then
$$y + (6 \times 4.25) = 42.5$$
$$y + 25.5 = 42.5$$
$$y = 17$$
Hourly cost is £17

EXERCISE 16.2

1
$$(x + 6x) + (5y - 5y) = 13 + 8$$
$$7x = 21$$
$$x = 3$$
$$3 + 5y = 13$$
$$5y = 10$$
$$y = 2$$

2
$$4x + 12y = 64$$
$$(4x - 4x) + (12y - 3y) = 64 - 37$$
$$9y = 27$$
$$y = 3$$
$$2x + 18 = 32$$
$$2x = 14$$
$$x = 7$$

3

$$12x + 3y = 144$$
$$(12x + 3x) + (3y - 3y) = 144 + 6$$
$$15x = 150$$
$$x = 10$$
$$40 + y = 48$$
$$y = 8$$

4 $\quad 3x + 2y = 79 \qquad (1)$
$\quad 2x + 5y = 115 \qquad (2)$
$(1) \times 2 \qquad 6x + 4y = 158$
$(2) \times 3 \qquad 6x + 15y = 345$
$(6x - 6x) + (15y - 4y) = 345 - 158$
$$11y = 187$$
$$y = 17$$
$$3x + 34 = 79$$
$$3x = 45$$
$$x = 15$$

5 $\quad 7x - 5y = 8 \qquad\qquad (1)$
$\quad 2x + 7y = 36 \qquad\qquad (2)$
$(1) \times 7 \qquad 49x - 35y = 56$
$(2) \times 5 \qquad 10x + 35y = 180$
$(10x - 49x) + (35y - 35y) = 180 + 56$
$$59x = 236$$
$$x = 4$$
$$8 + 7y = 36$$
$$7y = 28$$
$$y = 4$$

EXERCISE 16.3

1 Let the number of £5 notes be x and the number of £10 notes be y.

$$x + y = 24$$
$$5x + 10y = 195$$
$$10x + 10y = 240$$
$$(10x - 5x) + (10y - 10y) = 240 - 195$$
$$5x = 45$$
$$x = 9$$

Answer: 9 £5 notes.

2 Machine A $\quad x + 2y = 16$
Machine B $\quad 3x + 2y = 36$
$(3x - x) + (2y - 2y) = 36 - 16$
$$2x = 20$$
$$x = 10$$
$$10 + 2y = 16$$
$$2y = 6$$
$$y = 3$$

Answer: 10 of product x and 3 of product y.

3 Let the dress be x and the shirt be y.
$$4x + 2y = 1100$$
$$30x + 22y = 10\,000$$
$$44x + 22y = 12\,100$$
$$(44x - 30x) + (22y - 22y) = 12\,100 - 10\,000$$
$$14x = 2100$$
$$x = 150$$
$$(4 \times 150) + 2y = 1100$$
$$600 + 2y = 1100$$
$$2y = 500$$
$$y = 250$$

Answer: 150 dresses and 250 shirts were made.

4 Peter: $\quad 38x + 4y = 352 \quad (1)$
Matthew: $\quad 40x + 6y = 392 \quad (2)$
$(1) \times 3 \qquad 114x + 12y = 1056$
$(2) \times 2 \qquad 80x + 12y = 784$
$(114x - 80x) + (12y - 12y) = 1056 - 784$
$$34x = 272$$
$$x = 8$$
$$304 + 4y = 352$$
$$4y = 48$$
$$y = 12$$

Answer: basic pay is £8 per hour and overtime is £12 per hour.

EXERCISE 16.4

1 $\quad 4x - 2x > 9 - 5$
$$2x > 4$$
$$x > 2$$

2
$$6x + x \geq 18 - 11$$
$$7x \geq 7$$
$$x \geq 1$$

3
$$-2x - 3x > 7 - 22$$
$$-5x > -15$$
$$x < 3$$

4
$$3x - 2x \geq -5 - 19$$
$$x \geq -24$$

5
$$-6x < -9 - 25$$
$$-6x < -34$$
$$x > \frac{-34}{-6}$$
$$x > 5\frac{2}{3}$$

6 Let the number of bandages be x and the number of pressure pads be y.
$$2x + 5y < 120$$
$$2x + (5 \times 14) < 120$$
$$2x + 70 < 120$$
$$2x < 50$$
$$x < 25$$
Answer: 25 bandages.

7
$$x + 2y \leq 64$$
$$3x + y \leq 72$$
$$3x + 6y \leq 192$$
$$(3x - 3x) + (6y - y) \leq 192 - 72$$
$$5y \leq 120$$
$$y \leq 24$$
$$x + 48 \leq 64$$
$$x \leq 16$$
Answer: 16 widgets and 24 gadgets.